KV-317-344

BLACK OPS AND BEAVER BOMBING

BLACK OPS AND BEAVER BOMBING

Adventures with Britain's Wild Mammals

FIONA MATHEWS AND TIM KENDALL

A Oneworld Book

First published in Great Britain, the Republic of Ireland
and Australia by Oneworld Publications, 2023

A CIP record for this title is available from the British Library

ISBN 978-0-86154-556-8
eISBN 978-0-86154-557-5

Typeset by Hewer Text UK Ltd, Edinburgh
Printed and bound in Great Britain by Clays Ltd, Elcograf S.p.A.

Oneworld Publications
10 Bloomsbury Street
London WC1B 3SR
England

Somewhere around Chapter 5, Tim was diagnosed with bowel cancer.

This book is dedicated with gratitude to the NHS staff who treated him, as well as to the medical researchers, the teams of volunteers, the clinical trial participants, and all the fundraisers baking cakes and running marathons in fancy dress.

Without them, *Black Ops and Beaver Bombing* would have been a lot shorter.

Contents

Contents

Preface: I-Spy ix

1 What Did the Beaver Say to the Tree? 1

2 Crashing Boars 38

3 On the Trail of the Lonesome Pine Marten 77

4 Water Voles and Earth Hounds 118

5 Hanging Out with Greater Horseshoe Bats 156

6 Tiggywinkle Goes Rogue 193

7 Who Cares What Colour the Squirrels Are? 235

8 A Phocine Good Story: Saving Grey Seals 272

Afterword: Reintroducing Cats and Dogs 313

Appendix: Red List and Population Review 321

Acknowledgements 333

Notes 335

Index 347

PREFACE

I-Spy

What was the last wild mammal you saw? The chances are that it'll have been one of our fantastically successful imports: a brown rat, or maybe a grey squirrel. It won't have been a wildcat or a greater mouse-eared bat, and probably not a pine marten or a water vole or any number of other species whose populations have crashed over the past century.

Of course, there's no shortage of mammals in Britain. Cattle and sheep graze the hillsides, domestic cats patrol our gardens, dogs chase balls in the local park, and the most prominent mammal of all – *Homo sapiens* – has shaped every field, hedgerow, river and forest to the point where there's no wilderness left. Whereas Bulgaria has over half a million hectares of forest classified as 'undisturbed by man', Britain, at almost twice the size, has a grand total of zero. Inevitably, our native species have suffered under this domination. Of those that survive, many are hanging on in isolated pockets, their numbers declining and their environments deteriorating around them.

We know that different kinds of privilege – class, ethnicity, and so on – can determine the destinies of individuals and entire communities. Yet these are mere details compared with the privilege of being human. Other species exist only with our permission, and if the extensive list of man-made extinctions is anything to go by, that permission can be withdrawn at any time. By the same measure, we have it in our gift to rescue our fellow creatures and allow them space to flourish. The word 'wilderness' originally referred to land occupied only by wild animals; our task should be to return those habitats to them before they disappear altogether.

It's already too late for the wolf, lynx, brown bear, elk, root vole, and many other species, all of which survived the last Ice Age but not the human expansion that followed. Some may yet be reintroduced from other countries. Others, like the aurochs, won't be coming back. In this book, we focus on the ones that remain. With enough luck and patience, you can see them for yourselves, and that's exactly what we've tried to do. Each of our chapters is a quest for a mammal species that's on the brink. Armed with binoculars, a hot flask and, regrettably, an inexhaustible supply of dad jokes, we've travelled from Scotland to the Isles of Scilly in search of our elusive subjects. Delighting in what we find, and despairing for what we don't, we explain the urgent problems facing each species and what needs to happen if we're to save them.

Every so often, the Vincent Wildlife Trust appeals for people to submit records of polecats they've found dead on the road. This has proven a bit embarrassing for our daughters if there are friends in the car when we've stopped to inspect some roadkill. Their embarrassment became more acute a few years ago, when scientists started asking people to scoop up and send them any dead specimens so that they could carry out post-mortems.

Polecats are smelly even when they're alive. Leave them by the side of the road for a few days and they become, let's say, ripe. As for the mystery of where we store the bodies before we get the chance to pack them off . . . Don't ask what we've got in the freezer.

These are the kinds of highly sophisticated techniques that inform conservation work. In 2018, Fiona and a team of expert scientists from the Mammal Society published *A Review of the Population and Conservation Status of British Mammals.*[1] The 700-page doorstopper, commissioned by Natural England, Natural Resources Wales and Scottish Natural Heritage, was the first such census for more than twenty years. Using all the latest research from hundreds of scientific sources, as well as extensive roadkill reports and 1.5 million records of mammal sightings, it gave a detailed account of habitat preference, population status, distribution and future prospects for every land mammal species, whether native or not. So turning at random to the entry for stoat, we learn that they range over most of Britain but with patchier coverage in the north-west of Scotland, that their numbers have fallen by nine percent since 1995 to an estimated 399,000 (plus or minus quite a few), and that their decline is caused by a similar decline in prey species such as rabbits. On the other hand, the lesser white-toothed shrew, otherwise catchily known as the Scilly shrew, can only be found on – you guessed it – the Isles of Scilly. The 1995 census gave an estimate of 14,000 individuals, but we've no idea what's happened since then, because nobody has tried to find out.

That 2018 population review provided much of the evidence for the first ever Red List of British terrestrial mammals,[2] drawn up by Fiona with her colleague Colin Harrower for the Mammal Society and officially accepted by the various government agencies. The Red List tells us that a quarter of Britain's forty-six native land mammal species are currently threatened with extinction, and that a still larger proportion of our species – forty-four percent – is officially 'at risk'. On publication day, the media led with the story, and the hashtag #quarterofuk trended on Twitter for twenty-four hours. Reassuring though that measure of public interest may seem, the challenge is to know what to do when the phone stops ringing. The great legacy of the Red List, apart from the role it plays in raising awareness, is that it judges governments according to criteria established by the International Union for Conservation of Nature (IUCN) and accepted across the world. No longer is species conservation an African or Amazonian issue about which we can feel self-righteous from afar while letting our own wildlife disappear before our eyes. (Trophy-hunting for lions in Africa? *Totally immoral!* Shooting beavers that are inconveniencing landowners? *An unfortunate economic necessity.*) An assessment of Endangered for the water vole and red squirrel in Britain means that those species face the same degree of threat as, for example, the Asian elephant in India, the south-western black rhinoceros in South Africa, or the three-banded armadillo in Brazil. While we sit back to enjoy *Springwatch* or the latest David Attenborough documentary, we flatter ourselves that we're a nation of animal lovers. The Red List obliges us to prove it.*

---

* We doubt that anyone wants to read the 150 pages of IUCN guidance, so we've included the complete Red List for British mammals in an appendix, together with a mercifully brief summary of underlying principles.

Although our chapters feature the charismatic headline-grabbers, such as those species tumbling towards extinction (red squirrel, hedgehog), or the ones around which a public controversy rages (beaver, wild boar), we also make the case for the unloved and overlooked. When there's an opportunity for a guest appearance – like the Scilly shrew in our hedgehog chapter – we take full advantage. Shrews and voles matter, both for their own sake and because the ecosystem would collapse without them. The field vole is Britain's most abundant wild mammal, and doesn't even get close to inclusion on the Red List as a species of concern. Twenty years ago, there were more field voles than humans in Britain, but not any longer; the number has fallen from 75 million to 60 million. Well, what's fifteen million vanishing field voles between friends? The species spends its days in a network of tunnels in tussocky grass where it can nibble in peace, out of sight of aerial predators. Next time you're in some unkempt grassland, perhaps at the edge of a sports pitch or on a walk in the countryside, put your hand in the grass and feel for the tunnels. Make an opening and you'll often see small piles of greenish pellets and little heaps of sawn-off grass blades. No crisis – there are plenty of field voles to spare. And yet they're a food source for owls, kestrels, stoats and weasels, and, when their numbers are falling as at present, birds of prey fledge fewer chicks. If you want to save the barn owl, start by saving the field vole.

The animals are the heroes of this book. They deserve to be so much more than an occasion for solemn journeys of self-discovery. We'll spare you the moody sunsets. You definitely don't want to hear about our spiritual awakening as we lay under the stars in the Adirondacks. What matters is that, shamefully, when it comes to biodiversity intactness Britain ranks 189th out of 218 countries. We have an urgent duty to rescue and restore our native species, so let's start the job in Chapter 1, on a river not a million miles from our home in East Devon.

ONE

What Did the Beaver Say to the Tree?

'Nice to gnaw you!' It's the only clean beaver joke we can think of, as we make the half-hour car journey with our daughters to one of a small number of rivers in England where wild beavers are found. Devon is a hotspot for beavers, which is to say that it's home to two distinct populations: one on the River Tamar in the west of the county, and one on the River Otter in the east. The 2018 population review made a very conservative estimate of 168 beavers across Britain: 12 on the Otter, and 156 in Scotland. At that point the Tamar beavers didn't exist, at least not officially.

So on a warm summer evening in July 2020, with the first Covid lockdown easing but not yet lifted, we headed for one of the prettiest rivers in the country to watch beavers. If you were a beaver, you might choose to live here too. In a sonnet addressed to his 'native brook' – he was born upriver at Ottery St Mary – Samuel Taylor Coleridge remembered how as a youngster he had 'skimmed the smooth thin stone along thy breast, / Numbering its light leaps'. Well, that isn't his best poem, and skimming a stone along a breast seems reckless to say the least, but all his life Coleridge was nostalgic for what he called his 'careless' childhood beside the River Otter. And with good reason, considering the opium addiction and the unhappy marriage that came later.

Our destination was the venerable old village of Otterton, seven miles south of Ottery. Otterton Water Mill, now restored and fully operational, was mentioned in the Domesday Book and wasn't new even then. It's all picture-postcard and lovely, in a rather understated way. You can't buy beaver-themed keyrings and tea-towels here. You can't even find more than a fleeting mention of beavers on the mill's website. A solitary concession to wildlife tourists is a blackboard at its café, where people have written up the latest sightings. Gruffalo, hobbit and mountain lion, we were reliably informed, had been spotted recently. Also present were otters (in a flagrant example of nominative determinism), kingfishers, buzzards and – someone had written – 'beavers (upstream)'. That created something of a dilemma, as we had it on good authority that the beavers were downstream. Who was right? Beavers are crepuscular, by and large. They like the warm mellow light of golden hour, and they tend to move silently. You need to know where they're going to emerge, or you need to be lucky. Hurtling up and down a riverbank trying to glimpse one through the vegetation before the light fails is unlikely to work. We made our best guess and started downstream.

There are two sorts of beaver in Britain. The first sort – the legals – have filled in all the paperwork. And there's a *lot* of paperwork. If they've come from abroad, they'll be subjected to a lengthy quarantine. They'll undergo stringent veterinary inspections, including, in some cases, invasive procedures to assess the health of their livers. Finally, with the forbearance if not necessarily the full-throated approval of the relevant agencies, they'll be given amusing names like Sigourney or Jean-Claude or Chewbarka, and released – into a large enclosure.

Then there are the illegals. These bad boys don't care about your paperwork. They certainly don't care about your enclosures. They turn up unexpectedly and unexplainedly from who knows where. Actually, some people do know where, but they're not talking. Call it black ops, call it beaver bombing, it amounts to the same thing: the wrong sort of beaver has made its home in the wrong place. One day you walk along the riverbank and notice some curious gouges near the base of a willow, as if someone had taken a small axe to it and then given up out of boredom. The next day it's been felled, leaving a pencil-shaped stump, and you see twigs and branches lying oddly in the river; they seem to be fixed in position, immobile against the current. At this point, government organisations have been known to close their eyes and put their fingers in their ears.

There'd been beaver sightings in East Devon as long ago as 2008, although the Devon Wildlife Trust took some persuading that these weren't simply of misidentified Otter otters. Then in 2012 a sick male beaver was found by a roadside, and died shortly after. Perhaps the mystery had been solved, and he'd been the sole cause of all the reports. Or perhaps not. In 2014, a family of beavers – the mother and two or three kits – was caught on film for the first time. This only confirmed what locals already knew: signs of their presence, if not the animals themselves, are simply too easy to spot by anyone who knows about beavers. In fact, visibility was (and remains) their best defence. According to rumour, Defra's default setting was to have them culled in case they carried disease, although it hurriedly denied any such dastardly intentions after a short and bracing exposure to public opinion. The government had also hoped to play a get-out-of-jail-free card by counting chromosomes; the Eurasian beaver has forty-eight, and its North American cousin forty, the two species having diverged roughly 7.5 million years ago. The beavers on the River Otter had forty-eight and, as such, were a bona fide native species. So a compromise was proposed: the beavers would be captured and allowed to see out their days rehomed in an enclosure. It was fun while it lasted, but now the grown-ups would take over and make the problem go away.

But just as it's impossible to squeeze genies back into lamps, so the beavers were not easily going to be bottled up again. Wildlife enthusiasts had been privileged to experience what hadn't been witnessed for many centuries: free-living beavers flourishing on an English river. No one was prepared to give that up without a fight. And so the campaign group Save our Beavers was founded. Letters were written to MPs, stories started appearing in the local and national press, and, when a schoolgirl stood up at an otherwise acrimonious meeting and made an eloquent plea for beavers to be left in peace on her local river, the matter was all but settled. Even the landowner on whose banks they had made their homes was keen for them to stay. Faced with this overwhelming public support, the government enthusiastically welcomed the beavers' presence and rolled out a series of reintroductions across the country. We're joking: of course it didn't. It bought itself time by commissioning the five-year River Otter Beaver Trial, and prayed that, somehow, someone or something would turn up to sort everything out.

In one respect, the foot-dragging was justified. Beaver bombers get results, but only by taking risks with animal welfare and disease control. All wildlife reintroduction programmes worry – or *should* worry – about inadvertently letting loose a disease that could infect not only existing native populations but other species too, including humans. Covid is merely the latest and most devastating in a long line of zoonotic diseases, among them HIV, Rift Valley fever, SARS, West Nile virus and avian influenza. Like most mammals, beavers can harbour diseases that you'd rather not catch. For example, the chance of being set upon by a rabid beaver is infinitesimally small, but it's happened several times in the United States, where rabies circulates among quite a few wild species. Then there's 'beaver fever', *Giardia duodenalis*, a waterborne parasite that acquired its beguiling but unfair moniker in the early 1980s after a group of hikers in Banff National Park, Alberta, caught the infection from stream water. A nearby population of beavers also carried the disease, although no one knows whether the beavers introduced the parasite to the area or were themselves infected by the water. Contamination by human faeces is by far the most common source of *Giardia*, but we prefer to pass the blame, and, after all, 'human' doesn't rhyme with 'fever'. Through a microscope, the *Giardia* parasite bears an uncanny resemblance to a Pac-Man ghost, albeit with endearing flagella that it can wave. This is scant consolation if it gives you a nasty gastric upset; and worse, in some species, it proves fatal. The good news is that water monitoring in Britain has found no evidence of an increase in *Giardia* where beavers are present.

What most exercised the government about the beavers was the tapeworm *Echinococcus multilocularis*. This little charmer has been found in a small number of beavers exported from Bavaria to other European countries, including one in a captive collection in Britain. As a result, Bavarian beavers have been banned from entering Britain. The origin of the River Otter beavers is *officially* unknown – and if they've been tossed out of the back of a van on a moonless night they could be from anywhere – but at one stage most of the animals kept in British enclosures were sourced from Bavaria. *Echinococcus* is among the deadliest of all parasitic diseases in humans. The larvae cause severe liver damage, and by the time the infection is diagnosed it's often untreatable. The most recent official data from the European Centre for Disease Prevention and Control reports 114 human cases in the EU during 2020, mainly in Germany, Bulgaria and France. This is the lowest rate since screening began in 2007, with cases having resolutely failed to track the increases in European beaver numbers. It's never been reported in Britain.

Echinococcus multilocularis infects foxes and dogs. The eggs are shed in their faeces and enter the environment, where they're accidentally eaten by rodents such as voles (and, potentially, beavers) and develop into cysts. In the natural scheme of things, the voles would then be eaten by foxes, the larvae in the cysts would develop into tapeworms, the tapeworms would produce eggs, the foxes would excrete the eggs, and so the circle of life that moves us all would keep turning. Given that rodents don't excrete parasite eggs, humans can't be infected by accidentally encountering beaver poo, or by swimming in rivers where beavers live. In fact, were a tapeworm capable of offering an opinion, it would express the devout wish to avoid us altogether, because we're dead-end hosts; foxes and dogs don't generally eat people and, if they did, we'd have more immediate things to worry about than a tapeworm. It's *just about* possible to invent a plausible scenario for infection; a fox might gnaw on a dead beaver, then poo on a riverbank, and when the family of humans came to have a picnic at their favourite spot, they'd little know the terrors lurking in the undergrowth. A pet dog increases the likelihood, but it's overwhelmingly more likely to catch an infection by eating a mouse (dead or alive) during its summer holidays in continental Europe. Defra has concluded that the risk of the parasite becoming established in Britain through beaver reintroductions is tiny but not zero,[1] which explains why it felt the need to capture the beavers and give them veterinary assessments via a combination of ultrasound and keyhole exploration of their livers, before restoring them to the wild with colour-coded ear-tags and a clean bill of health. Incidentally, that's how the matriarch of the Otterton colony was given her underwhelming name; now thought to be approaching the grand old age of twenty, 'Yellow Tag' is still active on the Otter at time of writing.

When it comes to reintroductions, things have moved more quickly on the continent. At least twenty-four European countries have successfully reintroduced beavers, while others have welcomed their arrival across land borders. We have Norway to thank for the survival of the species in Scandinavia; in 1845, with their numbers fallen to a few hundred in the south of the country, beavers were given last-minute legal protection, and that population has helped to seed recovery across much of northern Europe. Norway's neighbours fared less well. The Swedes banned beaver hunting in the 1870s, several years after they'd driven beavers to extinction, while the Finns blundered in 1937 by accidentally introducing seven North American beavers; so successfully have they outcompeted the Eurasian beaver that they now outnumber them by three or four to one. The other great source of reintroduced beavers in Europe has been Bavaria, where an isolated population survived and started to thrive. As their numbers have grown, nearly a thousand of them have been translocated to new watercourses – in Croatia, Hungary, Romania, Spain, Belgium, and as far afield as Mongolia.

Even these recoveries seem modest compared with the astonishing turnaround in North America – a continent that, as Rachel Poliquin puts it, 'was built on the back of the beaver'.[2] Just at the point when populations of Eurasian beavers had been exploited to near-extinction, a vast and seemingly infinite new resource was discovered that could satisfy market demand. The Pilgrim Fathers paid all their debts in beaver fur; the Dutch bought Manhattan – and, in doing so, created what would become New York City – as much for its 7,000 beavers as for the island itself; the Dutch East India Company converted beavers into 'several tons of gold annually'; the so-called Beaver Wars, beginning in 1640 and lasting more than six decades, erupted when the Iroquois ran out of beavers to sell to Europeans and started campaigns to oust neighbouring tribes. Fur traders formed the vanguard of westward colonisation, eradicating beavers wherever they went. Before Columbus, there may have been as many as 400 million beavers from coast to coast. That number dwindled to around 100,000 as they were relentlessly hunted down and skinned.

If you were a gentleman looking for a stylish hat that wouldn't collapse at the first hint of rain, beaver was the fur for you. All those behatted figures in paintings from the Dutch Golden Age were wearing beaver. The puritanical sugar-loaves and cavalier wide-brims were beaver; so were the tricorne and bicorne hats of the Napoleonic era (Napoleon himself having always insisted on beaver fur); the Wellington, the Paris Beau, the colonial, and the top hat that became an icon of upper-class life were beaver. Whatever the latest fashion, until the mid-nineteenth century a quality hat would be made from beaver. Other animals were available, but their fur lacked the strange characteristic that made beaver so much more precious to the hatter. Separate the beaver's coarse guard hairs from the underfur (or fur-wool, as it's sometimes called), and you're left with an inch-thick layer of soft material with properties that lend themselves to felting. Each hair in the underfur is covered in microscopic barbs that, when abraded by means of pummelling and boiling, will mesh the fur together in a thick matt finish. Unlike most furs, beaver felt will keep its shape and is waterproof, making seamless hats that can withstand the storms of northern Europe. That quirk explains why hundreds of millions of beavers have been killed.

It was a close-run thing but, thankfully, beavers didn't go the way of the passenger pigeon: efforts to reintroduce and protect them began in Canada as early as the 1830s and, by the time they became a national symbol, their future was assured. In the USA, uneven distributions of beaver colonies were addressed by relocating them, although this effort was never quite so high-minded as it sounds, because the primary motivation was to guarantee 'fur for the future'. There's grainy footage of one project from 1948 that involved catching and crating seventy-six beavers and dropping them with parachutes out of aeroplanes – a literal beaver bombing – into the remote wilds of Idaho. Spare a thought for poor Geronimo, the stunt-beaver who endured countless aerial descents so that trappers could be certain that the box would open on impact. It's thanks to heroes like him and his pioneering peers that there are now twenty million beavers across North America.

Granted permission to reintroduce just one species of our choice, without hesitating we'd pick beavers. Whoever tells you that they aren't amazing is lying. They're the biggest prize because they transform landscapes. Reintroduce a large carnivore and, well, you have a large carnivore in your landscape, often with marginal benefits.[3] Reintroduce beavers and there are guaranteed massive gains across entire ecosystems. Water quality improves, flood risk is reduced, and a whole host of species increase in number: invertebrates, fish, amphibians, even other mammals such as water voles. Beavers are what ecologists call a keystone species, shaping the environment around themselves to *almost* everyone's advantage.

Beavers revolutionise the hydrology of their habitats. This process begins with the one thing that everyone knows about them: they build dams. At least, they build dams where the water level is too shallow (roughly speaking, less than a metre) to allow them to feel safe. Beavers out of water have been compared to a chicken nugget walking through the landscape, vulnerable to rapacious predators. Rivers are their refuge, and they design their safety by slowing the flow and creating deep pools into which they can dive when they feel threatened. The underwater entrances to their lodges provide extra security. Not for nothing has the eager beaver become a symbol of honest endeavour; dragging branches across land to the river is hard work, and beavers have only little legs. Over time, repeated use of the same trails moulds depressions. The depressions become larger, and, where the water table is high, they fill with water to form a network of canals. Beavers will also actively excavate canals, dumping silt in piles as they go. The resulting network, rich in biodiversity with its small channels radiating out from the main water body, makes the daily journeys of the beaver and its cargo of vegetation considerably easier and safer.

The size of their dams varies, but Eurasian beaver dams are generally quite modest affairs, and, in times of low rainfall, they'll also make tiny dams to keep the water in the network. No one will be shocked to hear that by comparison their American counterparts supersize everything. The materials of choice on both sides of the pond are branches, small twigs, mud and stones, but beavers will make use of whatever happens to be available: in 2016, canoeists in Wisconsin spotted a prosthetic leg in a beaver dam and returned it to its grateful owner. These feats provide defences against flooding. Beaver wetlands have been described as giant sponges, with scientists estimating that thirty percent of water entering beaver-populated land stays there rather than cascading downhill to deluge the nearest town.[4] A pair of beavers introduced into a strategically placed enclosure in north Essex are reportedly doing a much better job than previous human measures to protect a local village from recurrent flooding.

Imagine a future in which beavers across our river systems create natural flood defences, and the savings in money and human misery that ensue. Converting this into reality would require an upheaval in our attitudes. Over the past 500 years, and particularly in the last century, humans have tried to bend the course of rivers and streams to their wills. Channels have been moved to more convenient locations, and curves and kinks have been smoothed out. Our small local river, the Coly, is typical. Upstream, large field drains have converted marshy meadows and scraps of wet woodland into places more suitable for large machinery and intensive grazing. The remnants of an oxbow lake have been filled in to provide a field arrangement that makes cattle grazing easier. For that matter, the mammal that you're most likely to spot in the Coly is a cow, drinking or trying to cool down. (This contravenes multiple regulations that aren't ever enforced.) The ponds adjacent to the river, once teeming with tadpoles and fish fry, have also been filled in with soil and rubble; perhaps the farmer had a particular problem with liver-fluke, which lurks in mud-dwelling snails, or perhaps he thought he was tidying up. There's nothing unusual about any of this; three-quarters of Britain's ponds have been lost,[5] which is why in spring tadpoles can often be seen floundering in tiny puddles. Entire generations of farmers have been indoctrinated with the commandment to get water off their land as quickly as possible, with disastrous consequences for the environment. In the 2022 heatwave that left farmers fearing for their livestock (and the *Daily Mirror* fretting that Brits would soon face a shortage of chips), the National Farmers' Union took to national media to bemoan the fact that hundreds of thousands of gallons of water continued to escape into the sea, while the 'Infrastructure Tsar' called on the government to 'overrule NIMBYs' and build new reservoirs to combat drought. Beavers provide a softer, cuter, greener, and altogether furrier solution.

At the bottom of our garden, the Coly runs completely straight. Following a severe flood in 1968, it was rechannelled, and a smoothed floodplain was created, which from time to time does indeed flood. This stretch of river escaped much of the hard engineering that characterises flood defence schemes, but it's nevertheless walled on one side, with many large boulders strategically placed down the other, and is prone to extreme fluctuations in water levels. When it rains, the water races down to the sea several miles away, scouring the vegetation as it goes. Add to that the banks trampled by livestock, and the vegetation grazed flat up to the edge of the river, and you get a forbidding habitat for wildlife. There are no places for amphibians to spawn, ducks to nest, fish fry to grow, or water voles to take cover.

It doesn't have to be this way. As well as helping with flood control, the slower water flow in beaver habitats creates wetland mosaics that provide resilience to drought. Beavers can't fix climate change, but they're adept at dealing with the kinds of extreme weather events that are becoming increasingly common. In the USA, beaver wetlands are valued for their ability to halt wildfires, so decisions are now taken to introduce beavers to areas at risk. Besides anything else, it's a cheaper option: one dried-out Californian creek was transformed into rich biodiverse habitat at a cost of $58,000, instead of the $1–2 million estimate for heavy diggers to do the work. In the process of creating these waterscapes, beavers also clean their rivers. They make the equivalent of settling ponds, where sediment can drop out of the water. Sediment pollution – largely comprising topsoil from farmland that's swept into rivers after rain – is a growing problem that ravages biodiversity; each water-bill payer contributes to the clean-up costs. Again and again, monitoring experiments have shown vast improvements in the quality of water downstream of beavers.[6]

The creation of wetlands may be generally beneficial for the environment, but there are times and places when even the biggest beaver fans must admit it's unhelpful, such as when valuable crops such as seed potatoes are flooded. Beavers also have a propensity to block culverts and sewage outflow pipes. They're attracted by the noise of the running water and, hard-wired to act as if it's a leak in a dam, they seal up the hole with sticks and mud. The solution is either to move the beavers to a less problematic location or to install a 'beaver deceiver', thereby proving that there's no rhyme with 'beaver' that can't be exploited for comic effect. (The so-called 'castor master' performs a similar task.) Invented in North America, the beaver deceiver is a pipe that allows water to be drained from upstream of a beaver dam without the beaver noticing. There must be no noise from the pipe, otherwise the beaver will rush back to sort out what it perceives as a problem. These devices have already been used successfully in Scotland, as well as across the USA and Canada.

But what about the impact of beavers on fish stocks? Anglers distrust beavers. Or more accurately, some of the organisations that purport to speak for anglers distrust beavers. When a wild beaver population appeared on Tayside, the Angling Trust called for a thorough cull so that there was no risk of them reaching England. They carped vociferously about the beavers on the River Otter: these must be removed post-haste. Beavers may have once been native, the Trust's argument went, but riverscapes have changed completely since their extinction and are no longer suitable for supporting beaver populations. Without pausing to consider what that might reveal about the health of our rivers, the government minister George Eustice echoed their statement word for word. It was almost as if they'd compared notes. Back in the real world, most anglers love the natural environment and welcome the presence of beavers, but there are always one or two noisy eccentrics who insist that beavers are pesky piscivores busily depopulating rivers of their fish. Maybe they remember C.S. Lewis's Mr Beaver in *The Lion, the Witch and the Wardrobe*, who fetches dinner for his family by sitting next to an ice-hole and whipping out a number of 'beautiful trout' with his paw. Gently explain to them that beavers are herbivores and they'll assure you that they know better. What are those beaver dams for anyway, if not to trap fish?

Beavers on the River Otter have indeed attracted complaints from anglers, all fastidiously recorded in the official Trial's final report. Fly-fishing for trout on the Otter is a popular pastime, but one angler described how a 'beaver-felled tree had obstructed their ability to wade through the river', while another (or possibly the same angler having a bad day) grumbled that 'beaver tourists' along the riverbank were disturbing the fishing. Whether the God-given right of every Englishman to wade unobserved and unimpeded through the nation's waterways should trump the restoration of biodiversity probably isn't a moot point, especially at a time of ecological collapse. Take the River Otter catchment area, which – according to the project report – 'has depleted fish populations resulting from chronic diffuse pollution, poor habitat diversity and man-made barriers to fish migration'. That sounds bad, but it's also completely normal. The Environment Agency's 2020 report under the Water Framework Directive showed that only fourteen percent of rivers in England are of a 'good ecological standard', and not a single river or lake in the entire country is of 'good chemical standard', polluted as they all are with sewage, industrial discharge and agricultural run-off. While the government and a small number of anglers and landowners get hot under the collar at the very thought of beavers, they remain remarkably relaxed about the ongoing catastrophe over which they preside.

The hostility of the Angling Trust is particularly perverse, because results from the Otter and its tributaries confirm what has already been established time and again by surveys in North America and mainland Europe: the presence of beavers increases fish stocks. In the USA, beavers are being deliberately relocated to areas where salmon numbers are collapsing as part of a strategy to reverse the declines.[7] There are various reasons why this works, all related to the cleaning of the water and the creation of pools and shallows that provide the varying depths and the cooler temperatures needed by fish. The effect has been measured on a reach of one of the Otter's tributaries, the River Tale, where beavers have built dams. The total number of fish was more than a third greater in the areas around the dams, and trout were especially abundant; mature fish take advantage of the beaver ponds for shelter, while juveniles live in the riffle below. Numbers of minnow and lamprey have increased, the three-spined stickleback could be found nowhere on the river except in beaver areas, and only the bullhead, which likes fast-flowing water, doesn't care for this new arrangement.

The sole objection from the Angling Trust that stands up to scrutiny has to do with the ability of salmon and trout to migrate unimpeded. As a spokesperson for the Trust puts it, 'All fish species need to move around the river system in order to properly complete their life cycles.' The Otter itself provides a cautionary tale about the damage caused by obstacles to fish migration. The river was once famous for its salmon. In the 1800s the tidal stretches were heavily netted, and many landowners put out fish traps, the first of these being next to the weir just upstream of the mill. By the middle of the nineteenth century, these practices had so depleted the stock that all obstructions were banned, and in 1863 the river was again reported to be full of salmon. That revival was short-lived owing to heavy poaching and the reintroduction of obstacles; the Otter was again closed to migratory salmon and sea trout by 1888, with Otterton weir, positioned near the tidal limit, serving as the main barrier for fish entering the river. Over recent decades, obstacles have been removed, and the installation of passes that allow fish to circumvent remaining barriers has restored the migratory fish runs after an absence of more than a century. Amidst all this progress, introducing into the river a mammal that is renowned for its obstacle-constructing capacities may seem counterintuitive.

The extent of the problem can be overstated. After all, beavers, salmon and trout have evolved together and coexisted for millennia, without beaver activity driving migratory fish to extinction. On the Otter itself, where beavers haven't bothered to dam the main streams, there's no issue. On its tributary, the Tale, most dams are sufficiently modest that the fish can jump them, admittedly often making several attempts before they succeed. It's also possible for juveniles and smaller species to wriggle their way through the dam walls: the structures only *impede* the flow of water; if they stopped it altogether they'd soon give way under the pressure. Beaver-modified habitats aren't neat and orderly; there can be numerous channels, each of which will have dams in various stages of construction and collapse, providing opportunities for fish to find alternative routes. In those rare cases where the passage of fish is completely obstructed, there's the option to intervene and remove the dam. At the moment, the increase in trout numbers suggests that beaver dams are having only positive effects. When it comes to salmon, unfortunately the debate is hypothetical: salmon numbers collapsed to almost nothing before the beavers moved in, and won't recover until water quality is addressed. Oh, and beavers can help with that.

When they're not obliging anglers to step over fallen trees, the beavers on the River Otter have been remarkably well-behaved. This is, of course, entirely a product of their environment rather than their inherent goodness. If ever there were a location designed to minimise their impact on the landscape, this must be it. Somehow the trial report keeps an even tone when it fastidiously lists the cost of their damage to a maize field (£1.33) and a small tree in a rural garden (replaced for £18, including a stake and a tree guard). Beavers have no incentive to bother building dams on this part of the Otter because the water is slow-moving, with depths that provide natural protection against any passing wolves, lynx and bears (and, for that matter, gruffaloes). One bank is steep and thick with food in the form of trees and bushes. These beavers are unlikely to be cornered or taken by surprise, and they know it. With sufficient opportunity and inclination an otter might attack a beaver kit, but the biggest non-human threat is from dogs, which cross them at their peril: several years ago, there were lurid news reports that a beaver had 'had a proper go' at a dog that came too close, and would have killed a smaller breed. On another occasion, an observer recounted how a badger had come for a drink in the river, lost its footing on the steep bank, and slid down into the water with a splash. Typical badger. The nearest beaver didn't take kindly to the ensuing commotion, erupting out of nowhere to bite the badger on the nose. Beavers, of course, have teeth that can fell trees, so this must have hurt more than somewhat. There are photographs of the badger thrashing around. It can be hard to discern emotion in an animal's face. That particular badger looks absolutely terrified.

We've been beavering away to bring you the following factoids:

Beavers are the world's second-largest rodent, after the South American capybara. They can weigh up to thirty kilograms, which is two or three times heavier than a Eurasian otter. That's still nothing compared to *Castoroides*, a species of giant beaver that became extinct in North America at the end of the last Ice Age just over 10,000 years ago. They were two metres long, with six-inch teeth, and weighed as much as a black bear.

Beavers usually live for about ten to twelve years in the wild. In common with only three percent of mammal species, they pair for life, and DNA analysis shows that Eurasian beavers are more trustworthy than humans when it comes to monogamy. They mate in the water, and sex lasts only a couple of minutes. We'll leave you to make your own comparisons there.

Beavers give birth to one litter of two to four kits each year. These kits stay with their parents until they become sexually mature at about twenty months. Often they swim in a parent's wake, and hitch a ride by hanging on to the tail, a practice known as 'caravanning'.

Watch a beaver swimming, and you'll notice that it's almost entirely submerged. Sometimes the only part that breaks the surface is the very top of the head, on which sit, in a line, the nose, the eyes and the ears. The front paws are designed more for building dams than for paddling. Their epiglottis is located not in their throat but in their nasal cavity, and they can close their lips behind their incisors. These adaptations allow them to drag branches through a stream without swallowing water, and ensure that splinters don't go down their throat when they're gnawing wood.

On the subject of gnawing, the word 'rodent' comes from the Latin *rodere*, to gnaw. Beavers gnaw twigs and sticks like we eat cobs of corn, grabbing hold of both ends and rotating them with their paws.

A beaver's teeth are startlingly orange, owing to their iron-rich enamel. This evidently took the fancy of our early English ancestors, who valued them highly enough to go to the expense of mounting beaver incisors in elaborate gold settings to make jewellery.

A foot long and six inches wide, the tail is scaly and shaped like a paddle. It's a multi-tool device: beavers use it as a rudder, for temperature regulation, and to thump the surface of the water when they feel threatened. In the Renaissance, the scales provided convenient proof that the beaver was part beast and part fish, which meant that the tail could lawfully be eaten during Lent.

Beavers are territorial and will fight as a last resort, but they have ways of avoiding conflict, including tail-thumping. The most peculiar is a 'stick display', during which they'll stand on their hind legs in the shallows with a stick in their mouths, and move rapidly up and down, slapping the stick against the water.

Like rabbits and hares, beavers exhibit caecotrophy, which is a fancy way of saying that they eat their own poo. This gives them a second chance to digest nutrients that their body failed to extract first time through.

There are documented cases of beavers being killed when a tree that they're felling lands on them. There are also *undocumented* cases. When Fiona once enquired as to why there were fewer beavers in an enclosure than there should be, she was given that very convenient excuse. Everyone kept a straight face.

Beavers were already living in what we now call Britain almost a million years ago. Experts disagree over the exact date of their extinction, but it was some time in the last millennium. Michael Drayton in his interminable poem *Poly-Olbion* (1612) describes beavers as a 'now perish'd beast . . . to this isle [i.e. Britain] unknown'. They lasted longest in Scotland, where Hector Boece reported that they were still abundant on Loch

Ness in 1526; one lovely, if fanciful, suggestion made in recent times is that Nessie is or was a family of beavers. Things were bleaker south of the border, where they may already have vanished from England by the end of the eleventh century. There's a single account of a bounty paid for a 'bever head' in North Yorkshire as late as 1789, but that's more likely to be fraud or honest misidentification than an accurate record. We know that Thomas Cromwell imported four beavers from Danzig in the 1530s to add to his exotic menagerie, which also included an elk and an unidentified 'strange beast' for which he purchased a velvet collar before gifting it to the king. (Henry seems to have been unimpressed, if having Thomas executed within a year is anything to go by.) The historian Gerald of Wales recorded in 1191 that beavers were hanging on in the River Teifi – the only river in England or Wales still to have them, he claimed. 'Hanging on' is right, because Gerald credited beavers with complex teamwork in moving timber to rivers: 'Some of them, obeying the dictates of nature, receive on their bellies the logs of wood cut off by their associates, which they hold tight with their feet, and thus with transverse pieces placed in their mouths, are drawn along backwards, with their cargo, by other beavers, who fasten themselves with their teeth to the raft.'

This isn't even the most unlikely of Gerald's forays into natural history. He goes on to repeat one of the hoariest myths of all, found in such worthies as Aesop, Cicero, Juvenal and many others: that beavers are self-castrating. No prizes for guessing what this is a load of. It starts with the fallacy that beavers are hunted for their testicles, which they'll bite off when they find themselves cornered with no hope of escape. Gerald coyly observes that the beaver will 'ransom his body by the sacrifice of a part', throwing away the thing that is sought and therefore satisfying the hunter, who'll collect the prize and call off the pursuit. That 'part' isn't explicitly named, but Gerald leaves little to the imagination when he describes how a castrated beaver, if hunted again, will run to an elevated spot and lift up his leg to show that he's not worth the trouble. Gerald even engages in some dubious etymology to make his case: the Greek for beaver is *castor*, he says, because it castrates itself. Gerald is partially right. The Eurasian beaver is *Castor fiber*, which combines the Greek and Latin to translate as 'beaver beaver': the beaver is, to borrow the words of the song, so good they named it twice. None of this has anything to do with castration.

The first problem with Gerald's account is that beavers' testicles are internal. It's very hard to sex a beaver without massaging the skin to feel for the presence of a penis bone. (A word to the wise: this is best done when the beaver is under general anaesthetic, if at all.) True, you can poke around in a beaver's anal glands to encourage secretions, which, in the Eurasian male beaver, will be liquid and yellow, and in the female viscous and light grey, but we suspect that Gerald didn't do that. For a start, he didn't have the Pantone colour charts that are used by beaver experts nowadays. The second problem for Gerald is that beavers were – and still are – hunted not for their testicles, but for castoreum, a fatty secretion made by both males and females in sacs near the base of the tail. This substance would sound delicious if it were a wine and not the gloop from a beaver's back end: it's been described as 'an odorous combination of vanilla and raspberry with floral hints'.[8] Beavers use it for scent marking; when humans aren't smearing it onto traps to lure beavers, they add it as a food flavouring and as a tincture in various perfumes and cigarettes, which is one of many reasons why you should always read the label. (That's no excuse to stop taking your castor oil, which has nothing to do with beavers.) Beavers can be milked for castoreum, which sounds grim enough, but killing them is quicker.

So for many centuries, until the past decade or two, beavers were absent from British rivers. The latest Red List for Britain's mammals therefore records a startling if still modest and precarious comeback: it classifies beavers as Endangered in Scotland, Critically Endangered in England, and Not Assessed in Wales. Each of those verdicts is the product of a complex history. One of the peculiarities about a successful reintroduction is that the species immediately becomes Critically Endangered; this sounds serious but is preferable to Extinct.

Scotland appears to be the best-performing of the British nations, its reintroduced beavers having already graduated from Critically Endangered to merely Endangered. If numbers reach more than a thousand mature individuals and populations become established in at least five locations, the beaver will be removed from Scotland's Red List altogether. Scotland currently has only two discrete populations, and their different history and legal status encapsulate the problems that have beset beaver reintroductions from the beginning. The population on Tayside has been there for well over a decade, having been founded or supplemented (depending on whom you believe) by escapees from an enclosure on the Bamff Estate in Perthshire. It's breeding well – too well according to some local landowners, who at one point madly took it upon themselves to cut down their own trees in order to dissuade beavers from moving in. Before 2019, it was perfectly legal to shoot beavers. They've since acquired a fig leaf of protection, so licences are obligatory. Most have been granted without so much as a conversation about other options. The statutory nature conservation body for Scotland, NatureScot (previously Scottish Natural Heritage – it rebranded itself, having got fed up with people thinking it looked after castles), reports that in 2019 eighty-seven Tayside beavers were killed under licence because they were supposedly causing problems of one kind or another. In 2020, 115 were killed, and another 87 in 2021. Shooting is meant to be a last resort, implemented only where mitigation has failed and translocation is unfeasible. Scotland seems to encounter a lot of last resorts.

The other beaver colony is at Knapdale in Argyllshire, the home of the Scottish Beaver Trial, where sixteen animals sourced from Norway were released in 2011. The publicity claims that the trial has introduced the first wild beavers in Scotland for 400 years, but in fact it post-dates the Tayside beavers and may even have been established as a response: if governments are able to demonstrate that a reputable scientific study is taking place elsewhere, they can ignore demands for wild populations to be given official status. The Knapdale beavers are isolated, their site having been chosen because its geography ensured 'good natural containment'. In other words, the beavers are hemmed in by their landscape. They need fresh water, so they'd be highly unlikely to survive any attempt to disperse out of Argyll via the sea lochs or the Sound of Jura. The glacial topography of high rocky ridges ('knaps') and steep valleys ('dales') has also made it difficult for the introduced beavers to move very far, and, at over seventy percent, juvenile mortality has been extraordinarily high. The long story short is that beavers themselves would never have chosen this location. A cynic might claim that the study was designed to fail, although we couldn't possibly comment. Ironically, the Knapdale colony is now being supplemented with beavers from Tayside – which is, if nothing else, a better option than shooting them.

For all its problems, at least Scotland has beavers. There's even a full-time beaver specialist at NatureScot working to resolve problems wherever they occur. The story in Wales, where the beaver's official Red List status is Not Assessed, is the saddest of them all. The beaver in Wales has been caught out by a technicality. The IUCN has proposed the year 1500 as the starting date for Red List assessments – not for any particular reason except that it's easy to remember and a long time ago. Had there been proof of beavers in Wales since then, they'd be classified as Extinct, and this would bring with it some obligation to consider their reintroduction. However, Gerald of Wales's sorry tale about the last beavers on the River Teifi in 1191 doesn't inspire confidence that they persisted until 1500. Despite an oral tradition of beavers having survived in North Wales until the seventeenth century, the formal records offer no supporting evidence. Small wonder that they died out so soon when you take into account the value placed on their fur. The tenth-century King Hywel Dda introduced a lucrative law that beaver skins were reserved for royalty. They were worth five times as much as marten skins, ten times as much as stag and ox hides, and about the same as 'a good horse'. The only surprise is that beavers lasted as long as they did. So it's random bad luck that the beaver in Wales is Not Assessed, bringing no pressure on anyone to consider reintroducing them. They're just going to have to reintroduce themselves; if rumour and trail camera footage are to be believed, this has already happened on the River Wye.

Meanwhile in England it took a fluke to save the species. When Fiona and her colleagues were writing their 2018 population review on which the Red List came to be based, the beavers on the River Otter were treated as part of a trial. For that reason, they couldn't be included as post-1500 evidence of a free-living and breeding population. Although occasional records existed of beavers elsewhere, there was no official recognition that any of those were established families rather than solitary non-breeding individuals. Some escapees had taken up residence at the Cotswold Water Park, but they were known to have been sterilised before their bid for freedom, so no luck there. There were also beavers in Kent, which, coincidentally, Fiona had first encountered when they were being quarantined after their arrival from Norway and Bavaria. In captivity, they'd developed a serious obsession with carrots, so one of Fiona's students assessed whether they would go back to a more natural diet after being released into their 53-hectare enclosure, which indeed they did. As seems always to be the way with beavers and enclosures, several of them subsequently escaped. Not much effort has been put into finding out whether there's now a breeding population at large, despite regular sightings. The people involved in the original project are in no hurry to draw attention to their leaky enclosure, nor is Natural England desperate to repeat the problems that it encountered on the River Otter. So the Kent population didn't help the beavers' Red List status: the rule of thumb seems to be that they don't count if they haven't been counted. It was left to Devon's other beaver population, on the Tamar and its catchment area, to swim to the rescue.

The River Wolf is tiny: a tributary of a tributary of the Tamar. (Otter, Wolf – why can't the beavers get their own river?) Its claim to fame, for our purposes, is that it's the most heavily beaver-impacted stretch of water in Britain. The beavers on the Otter leave virtually no trace. The same can't be said here. One evening several years ago, a beaver dam on the Wolf collapsed, causing a sudden and sizeable wave. The beavers got to work immediately; finding the water pressure too strong to put matters right, they set about creating a side-channel that diverted the flow sufficiently to enable them to rebuild the main dam. As that particular incident may suggest, the beavers on the Wolf haven't kept a low profile. Evidence for their presence is everywhere. In such a minor river, to make pools that are sufficiently deep beavers need to build dams, and lots of them. Picture every possible sign of beaver activity, pushed to extremes of caricature: broken stubs of trees, and others half-gnawed, cartoon-style, with a pile of fresh woodchips at their foot; trunks left where they fell as if wantonly destroyed like hens in a fox attack, or dragged in desultory fashion half into the river before the beaver had a better idea and went off to do that instead; thick dams twisted together out of rushes and branches, with a number of sawn wooden planks all too prominent; deep pools with sunken trees standing tall and incongruous in the middle of them. The Wolf empties into Roadford Lake, a huge reservoir supplying North Devon as well as Plymouth and the south-west of the county; when control centre alarms were triggered by rising water levels, the culprit was found to be a beaver that had built its dam right next to a water meter. Ecologists reported how difficult it was to dismantle the dam: so tightly enmeshed were the materials that there was no obvious weakness on which to start tugging.

For some time, officials stuck determinedly to the position that these were solitary individuals from the Tamar. A beaver hoicked out of a slurry pit by the RSPCA in 2012 must have been one of a group that had escaped from an enclosure twenty miles upriver four years earlier – mustn't it? (In that particular incident, the story goes that three beavers made their break for freedom after a badger dug under their enclosure.) Faced with mounting evidence, no wishful thinking could last for very long. Nor could any scheme to remove the beavers. One couple running a bed-and-breakfast had set up hides along the river that flowed through their land, and made a very good supplementary income out of their beaver tours. When officials suggested to them that the beavers should be captured and removed, they calmly replied that their 11,000 Twitter followers would be fascinated to hear about those plans. The officials went away and didn't come back. There was, however, one thing that everyone could agree on: after centuries of absence, England once again had a population of free-living beavers. For Red List purposes, they had to be assessed.

In early August 2020, the government announced its response to the completion of the River Otter Beaver Trial on the other side of Devon. With no significant opposition, the beavers were given permission to stay. It had taken an extraordinary run of events for the illegals to be granted residency. Accident and good fortune had played their part, as had the nimble footwork of a dedicated team of organisations and volunteers, including the Devon Wildlife Trust, the universities of Exeter and Southampton, Clinton Devon Estates, and Derek Gow Consultancy. Thanks to each of them, the unthinkable had become the unstoppable. Beavers may now go forth and multiply – not that they hadn't been doing that already, blissfully unaware of the arguments raging around them. In 2015, there were two known breeding pairs on the Otter; five years later, and with the addition of further individuals to broaden the gene pool, there were at least twenty. During the trial, any peripatetic beavers were captured and brought back to the Otter. Now they're left to spread naturally. Telltale tooth marks reveal that they've made it at least as far as the middle of Honiton, and debate rages about whether natural dispersal or beaver bombing could explain the presence of beavers on an isolated pond in the gardens of the National Trust's genteel Knightshayes in mid-Devon. As Devonians ourselves, we look forward to the day – not very many years away – when those two pioneering populations on the Otter and the Tamar will meet and cover the entire county.

Predictably, the government has announced yet another consultation period, tiptoeing towards a policy decision that beavers must be introduced 'cautiously', and stalling wildlife trusts up and down the country that have been storing Tayside beavers and are impatient to reintroduce them into their areas. All the well-meaning soundbites aren't worth the paper they're not written on. In fact, the consultation proposals laid bare the government's nefarious scheme: they don't want to reintroduce beavers, but they're sneaky enough not to want to be seen stopping the reintroductions either. To achieve this nuanced passivity they intend to price everyone out of the market: each beaver project will require a five- or ten-year plan backed up with feasibility studies and suitable funding sources, a local beaver officer, and a steering group that will be financially liable for the officer's salary and the costs associated with managing any negative impacts. Still fancy putting your name forward? There's no matching proposal to reward steering groups for any positive outcomes such as flood control or improved water quality. Oddly enough, the release each year in Britain of 57 million non-native game birds, most of them imported from European factory farms, faces none of these bureaucratic hurdles, despite the animals being classified in law as 'species that imperil UK wildlife'.

At least on the legislative front, the status of beavers in England seems to be improving. For a time, the government did its best impersonation of Paul Whitehouse's *Fast Show* character Indecisive Dave, the chap who's forever changing his mind because he agrees wholeheartedly with the last opinion he's heard. After two U-turns in quick succession – one prompted by the NFU and the second by the subsequent public outcry – they've now brought England in line with Scotland and listed beavers on Annex II of the Habitats Directive, making it an offence to intentionally capture, kill, disturb or injure them, or damage where they breed and rest. Their Red List status means that they also gain protection under the Wildlife and Countryside Act. About time too.

Recently, beavers seem to have become the must-have accessory for landowners who fancy dabbling in a bit of rewilding and can afford the massive enclosures. Answering the perennial question, 'What do you get a man who has everything?', in 2020 Boris Johnson bought his father Stanley an eightieth birthday present of some beavers to keep on his Exmoor estate. It's just about possible for us riff-raff to be cheered by this beavery enthusiasm while fretting over the risks for animal welfare; too many translocations end in fatalities. As the government meets a moveable object by doing its damnedest to turn itself into a resistible force, beavers languish in enclosures and lose their fitness for survival in the wild. A male that escaped from the Knepp estate in West Sussex lasted only a month before succumbing to septicaemia; two of the three beavers illegally released on the River Beauly died shortly after recapture; a beaver that fled its enclosure in Plymouth after severe floods was killed on a road that same night. Reports tell of long-confined beavers turning turtle and floating belly-up, quite dead, as soon as they're released. There are so many disadvantages to taking things slowly that it's tempting to sympathise with the Belgian ecologist who, frustrated by governmental bureaucracy, acquired 100 beavers from Bavaria and dumped them, all at once, across the Ardennes in the middle of the night. The occasional fatality is considered a price worth paying for the immediate conservation victory.

From the point of view of officialdom, beaver bombing does serve one useful purpose: it shifts liability. In 2013, a Welsh fish farmer sued the Environment Agency, complaining that he'd been 'literally eaten out of house and home' ('literally' is a great touch) after the Agency installed holts nearby to encourage otters to breed.[9] He lost his case, but reintroduction is a step up from encouragement. If a government agency gives the go-ahead for beavers, and then a crop gets flooded as a consequence of beaver activity, who pays? The determination to make volunteer steering groups financially liable amounts to a cynical buck-passing exercise, because proper government funding is the essential lubrication required to ease past these challenges. A simple compensation scheme of the sort used during the River Otter Beaver Trial – a tree gets damaged so you receive a cheque for its replacement – comes with the disadvantage that it turns beavers into a problem with a bill attached. These schemes are also difficult to manage; international experience with species ranging from wolves to elephants shows that claims of damage mysteriously skyrocket as soon as compensation is made available. So why not redirect money from some of the agricultural subsidies that actively harm our environment? A presence/absence survey, with an annual subsidy rewarding anyone with beavers on their land, is the best investment that government can make; it turns beavers into a valued asset, improves biodiversity and river health, reduces flood damage and risk of wildfire, and protects farmers who'll occasionally bear the brunt of their activities. Regardless, beavers are continuing to show up in all sorts of places whether the government likes it or not – and their appearance delights the locals. There are now thought to be fifty beavers on the Bristol Avon and River Frome, and a Twitter video recently went viral of a beaver spotted near Blandford Forum: 'Oh my God, it's a platypus! It literally is a platypus; Mum, that's an actual platypus!' The beaver's return to our rivers is the greatest wildlife story of the century, so let's celebrate it. We've had 'Free the Nipple'. The time has come to Free the Beaver.

From Otterton Mill, it's only a couple of miles to the sea, and a well-used path runs alongside the river. We met a few dog-walkers, and asked if they'd spotted anything. Yes, a bachelor party of mallards. Now, we like mallards too, but we weren't stopping for them. We meandered along, periodically using the clearings in vegetation to get closer to the river's edge. There was no breeze; the air was rank with the stench of Himalayan balsam, a sickly-sweet smear at the back of the throat. Beavers eat Himalayan balsam, but sadly not in sufficient quantities to halt the invasion. More pleasing were the daubs of yellow tansy, from which fairies make buttons – or so the pixies have told us. Suddenly a flash of iridescent blue, then another, and another: three young kingfishers took a brief rest on an overhanging branch before darting low downriver, emitting the occasional peep as they passed. It was a brilliant spot for kingfishers, with numerous nesting holes in the red sandstone bank. The combination of shallow waters and deeper pools, together with an abundance of perches, made for good hunting territories. Plenty of trout were jumping. Still no beavers, but we started to notice signs of their activity: stands of cut willow, with characteristic tooth marks.

And there was the place that we'd been told to look for: an open patch of grass, and a branch in the water where debris had accumulated. So we hunkered down on the bank, just the four of us, patient but alert. It was that time of evening when everything starts to settle, and noises seem magnified. The beavers, if they were going to appear at all, wouldn't be long now. There was some rustling to our right: not a beaver, not an otter, not even an early fox or badger. A young woman with a moon-sized camera lens and full sniper-style camouflage gear was inching through the undergrowth five metres away. Now voices were coming closer; they belonged to a couple in yellow hi-vis hiking jackets, who stopped and stood on the bank behind us. The camouflaged photographer tutted audibly – not, we suspected, because of any concerns about social distancing.

Our younger daughter saw it first. Lifting its nostrils fractionally above the water, an adult beaver swam past, no more than ten metres away. One moment there'd been nothing, then out of nowhere the beaver was in full view, aware of our presence and utterly indifferent. It was the celebrity, and we were just the paparazzi. It disappeared under some overhanging foliage and emerged clambering onto the far bank a minute or two later, conspicuously clumsier on land. The issue now was knowing where to look, because a little further off another adult beaver had briefly surfaced, accompanied by a kit. Not to be outdone, our first beaver came floating past with a stick of willow in its mouth. It disappeared behind the overhang, where we could hear it chomping on its meal. Soon enough it set off again, unhurried. What was it doing? Tim went to have a look, and almost walked straight into it as it stood upright on its hind legs on the bank eating more willow. It slid into the water, nonchalant as ever, and drifted back downstream. In the meantime, the other adult had climbed out onto an island with a leafy willow branch. The branch snagged on something underwater and drifted away. After a low-speed pursuit the beaver finally caught up with its dinner, and dragged it back against the current. The beaver happened to be surrounded by willow, so why that particular branch was worth the effort is an abiding mystery.

Most nights you'll sit for hours by a riverbank, getting cold and no doubt wet and seeing nothing, all the while knowing that the moment you give up and go home the beavers will start their party, complete with stick displays and dam-building competitions. But as seasoned wildlife-watchers will know, it's the routine failures that make the successes all the more glorious. The beavers were on form: the adults swam obligingly up and down several times, while two kits kept tight to the far bank, not yet as bold as their parents. Our eyes adjusted to the light as it deteriorated, but in the end even we had to admit that there wasn't much more to be seen. The kits and one adult had already gone. We were just about to give up for the evening when a flock of Canada geese came low overhead, honking. The remaining adult slapped the water loudly with its tail, and dived.

TWO

Crashing Boars

Many years ago, long before we turned up as undergraduates, our college appointed a new dean. He brought with him boundless charisma, a fully costed step-by-step plan for world domination – and a dog called Argos. Argos, the dean was often at pains to point out, was named after Odysseus' loyal hound, and had absolutely nothing at all to do with a popular retail outlet. His name wasn't even the biggest of Argos' problems. The college's ancient statutes decreed that a cat was the only domestic animal allowed on the premises, and the governing body of that august institution was disinclined to tinker with the wisdom of the ages. Thankfully, after long and agonised debate, a solution was found. Argos the dog may have looked like a dog, barked like a dog and cocked his leg like a dog, but Argos the dog was a cat.

We tell this story because it predicts the response of various British government agencies to the return of wild boar. There are, of course, crucial differences: Argos the cat lived happily ever after, unaware of those feline slurs, whereas wild boar face systematic campaigns to exterminate them. Nevertheless, both cases demonstrate the power of officialdom to oblige us to disbelieve our own eyes. Wander through a forest in Scotland today and there's a chance that you'll spot what looks for all the world like a wild boar. You'll be very much mistaken. As an endangered native species, wild boar would have a strong case for being afforded a number of legal protections, which is exactly the reason why the animal you saw is a feral pig. In the looking-glass world of Scottish law, a wild boar is only a wild boar when it's not wild.[1] The situation is hardly less curious in England and Wales, both home to what are subtly called 'feral' wild boar, the word conjuring up images of lawless youths wrecking the swings at the local playpark. No matter how many generations of wild breeding have passed, these are apparently not, then, your *echt* wild boar; these are escaped pigs with a bad attitude and an overdue date with a sausage machine.

When Fiona and her collaborators published their 2018 population review, their estimates for each native species were accepted by the devolved authorities as the basis for the various national Red Lists, with one major exception. The review had classified wild boar as Vulnerable in England and Scotland, and Endangered in Wales. Those verdicts were never going to survive official scrutiny. Having whispered behind their hands for a while, the three nations spoke with one voice. Boar should be Not Assessed; they'd been hunted to extinction before 1500, and could therefore be ignored. In the absence of reliable surveys, the mammal biologists had given a rough population estimate of 2,600 adult wild boar in Britain, and the politicians had overruled them: lifting the telescope to their blind eye, they announced that there were no wild boar anywhere to be seen. After further negotiation, this absurdity gave way to something fractionally more nuanced: although the Mammal Society's own publications would retain the scientists' verdicts, wild boar would be classed as Data Deficient in the official government-sanctioned documents. Perhaps, after all, they might be out there somewhere – but who knows where they came from, or quite how wild these so-called wild boar actually are?

The IUCN disapproves of the Data Deficient category, having foreseen how convenient it would be for politicians to park any problematic species there. If data really are deficient, national governments are expected to go and establish the facts for themselves. We're shocked – shocked! – to have to report that the governments of Scotland, England and Wales have done the square root of nothing. What's more, they're unlikely to be roused into action any time soon; better a quiet life than to spend time and money on a troublesome species. Were boar somehow to barge their way onto the Red List as Vulnerable or Endangered, our governments would come under pressure to protect them, and perhaps even to expand their ranges. Worse than that, they'd no longer be able to lecture impoverished African nations, because 'Do as we say, not as we do' is never the most beguiling of messages. When a Zimbabwean smallholder, growing barely enough to feed his undernourished children, has his crops destroyed by a herd of elephants, we expect those elephants not to be harmed. It's not a good look if, in the meantime, we're busily wiping out an endangered native species because we don't like it when they uproot a few bluebells or damage a cricket pitch.

A neat segue brings us to Yorkley in the Forest of Dean. Here, in 2015, village cricket came to a sudden halt when boar stopped play by digging up the outfield. Two years and one £15,000 fence later, it was déjà vu all over again after someone left a gate open. As an elderly club trustee put it, 'We were going all the way through the war but we just can't cope with this.' Six miles further north, Cinderford Town Football Club has a similarly distressing tale to tell. The team went several seasons without being able to play a home game; boar regularly shouldered their way through the fencing and rotavated the pitch. It's fair to say that such shenanigans have not always endeared them to the locals. We wanted to see these rascally oinkers for ourselves, so in October 2020, days before the second lockdown, we packed our wellies, raincoats, torches, binoculars and younger daughter, and headed north to Gloucestershire.

Boar had turned up in the Forest of Dean in 1999, after an unspecified number of them escaped from a farm near Ross-on-Wye. The location was strangely appropriate: the area had once been renowned for its boar. Henry III's Christmas banquet in 1251 sourced 200 wild boar from the Forest, a rate of attrition that couldn't be sustained. His grandson, Edward II, put in a similar order sixty years later, but by then for some reason there seem to have been no boar left. They clung on a while longer in Scotland, and were sporadically reintroduced (and again eradicated) on hunting estates through the centuries, but the boar that recolonised the Forest of Dean were the first to set trotter there for more than 700 years.

Boar farms started to arrive in England in the early 1980s, answering a growing demand for a meat richer and less fatty than pork and a higher standard of animal welfare. Who could possibly have predicted that wild breeding populations would spring up almost immediately? By the end of the century, there were at least two colonies in England. The older of these, on the Kent–Sussex border, owed its existence to the Great Storm of October 1987, when falling trees smashed holes in farm fencing; the other, in the west of Dorset, was founded after yet another farm escape in 1996. The simple rule with boar, as with beavers, is that if you keep them inside an enclosure, sooner or later they'll appear, abracadabra, *outside* that enclosure. (Yorkley Cricket Club and Cinderford Town will testify that the reverse is also true.) They prefer deciduous woodland, so with a vast area of suitable habitat to expand across, the Forest of Dean boar soon established themselves as the largest population in Britain.

Their cause was helped, and their gene pool diversified, in 2004 when more animals were dumped in mysterious circumstances on the western edge of the Forest near Staunton. Initially, the Forestry Commission reported that about a dozen had been released, and then, like all the best anglers' tales, the size grew in the retelling – first to forty, then sixty, and now seventy. In the absence of ear-tags their origins were unclear, but this was no jailbreak. Legend has it that a jilted Welsh farmer was to blame, taking revenge by depositing the boar on his ex's land. If so, he must have been out of his mind with rage, because his weirdly oblique strategy came at quite a cost: a healthy adult boar's carcass might fetch well in excess of £200.

Since then, the Forest of Dean has become the radioactive centre for debates about the desirability of reintroducing boar to Britain. Excitable headlines in the local press like to emote about dangerous encounters between boar and dogs; poachers bring guns into the Forest under cover of darkness; lawns are routinely destroyed, with no restitution for angry gardeners. Some residents can be fearful for their well-being: hearing the stories and seeing the damage, they imagine a threat far more severe than the reality. Pubs in the area develop reputations as pro- or anti-, and arguments flare over what the right number of boar should be. (The answer, according to a purple-faced minority, is zero.) Even the local organisation Friends of the Boar disbanded in acrimony, having split between the welfarists who believe no boar should be harmed under any circumstances, and the conservationists who are willing to sacrifice individuals if it means ensuring the long-term survival of the species. In the eye of the storm sits the Forestry Commission, a government department that, under the auspices of Defra, manages the Forest on behalf of the Crown. Whether fairly or not, what everyone seems able to agree is that they've botched the whole thing from start to finish.

So, anyway, are we talking about the return of a long-extinct native species, or just a gang of farm animals on the wrong side of the fence? Such purity tests have their uses if you're a politician looking for loopholes, but in scientific terms they're virtually meaningless. A boar is a pig is a boar. They're the same species – as their Latin names *Sus scrofa* and *Sus scrofa domesticus* make abundantly clear – and they fill the same ecological niche. Humans have always been fascinated by boar and their relatives: the earliest surviving piece of cave art, from at least 45,000 years ago, is of a Sulawesi wild pig, and 10,000 years later one of our ancestors went to the effort of depicting a wild boar on a cave ceiling in Altamira, Spain. Rather charmingly, the boar has eight legs, as the artist attempts to convey that the animal is in motion. It took a lot longer for us to work out how to tame a boar. We first domesticated the species in Eastern Anatolia around 8,300 years ago – not even an eye-blink in evolutionary terms. Then as farming spread westward into Europe, pigs inevitably bred with wild populations. So common were these interactions that we invented a word for their results: 'hybrid' comes from the Latin *hibrida*, which originally referred to the offspring of a wild boar and a domestic sow. Countless generations of farmers have deliberately cross-bred wild boar with domestic stock because it increases litter size and makes for more frequent farrowing.

In other words, what we call pigs are domesticated wild boar bred intensively and selectively for certain favourable characteristics, but then topped up with good old-fashioned wild boar genes when they stray too far from the template. And what we call wild boar is not some genetically uncontaminated free-living species that has never met a pig, but the product of countless generational interactions between wild and farmed animals. Love will always find a way. Consider, for example, the opportunities for romance provided by the practice of pannage, whereby farmers across medieval Europe (and, for that matter, in the New Forest even today) let loose their pigs into woodland every autumn to feed on fallen nuts. It's hardly surprising, then, that boar populations should carry the genetic markers of domestic pigs.

There are a lot of wild boar subspecies, depending on your definitions: some taxonomists say four, some twenty-five, and

Wikipedia currently plumps for sixteen. These are dotted across a native range that stretches from Japan to Portugal, and the south of Scandinavia to the north of Africa. (One reason why they never made it past the Sahara is that pigs *don't*, in fact, sweat like pigs; having no sweat glands, they share with Prince Andrew an inability to sweat at all, hence the need to wallow.) As wild boar farms become increasingly popular throughout the world, so escapees will introduce far-flung genetic markers into local populations. Farmed boar are themselves crossed with pigs by farmers hoping to give them a more placid temperament. Occasionally, someone will cunningly claim that only wild boar of the right subspecies should be considered native to Britain. 'Aha!', they'll cry, 'the Forest of Dean boar carry a genetic marker indicating Japanese heritage!'[2] This only tells us that, at some stage, the boar bred with pigs that had Far Eastern ancestors, not that some deranged rewilder secretly parachuted a payload of boar into the Forest from Japan.

For many conservation projects, reintroduced animals should be matched as closely as possible in genetic terms to those that have been lost; they can carry adaptations to the local environment that would be absent in populations sourced from elsewhere. This is a major concern for the potential reintroduction of European lynx to Britain, because the northern subspecies *Lynx lynx lynx*, found in Scandinavia and the Baltic, is likely to be better suited to our habitat and climate than the Balkan or Carpathian subspecies. In the case of boar, the argument is nonsensical, because of the genetic soup that interbreeding with domestic pigs over thousands of years has created. Luckily, the IUCN addresses this very situation:

> In some cases the original species or subspecies may have become extinct both in the wild and in captivity; a similar, related species or subspecies can be substituted as an ecological replacement, provided the substitution is based on objective criteria such as phylogenetic closeness, similarity in appearance, ecology and behaviour to the extinct form.[3]

The Forest of Dean boar satisfy these criteria: they look and behave like wild boar and perform the same ecological function. If there were any maladaptive traits derived from their genetic heritage – whether Japanese or otherwise – these would soon disappear under strong evolutionary selection pressures.

The final straw-grasping accusation levelled at the boar is that, owing to some unspecified aspect of their heritage, they'll breed faster than an imaginary 'native' boar. Even if that's true, so what? Using data from Italy and the Forest of Dean, Defra's Animal and Plant Health Agency shows that annual culling rates of 20–40% keep populations stable or slightly declining. By happy coincidence, the Forestry Commission culls at those rates, creating a sustainable harvest. Even if culling were reduced or stopped altogether, the boar population would stabilise within five years, albeit at a higher level. Densities become self-regulating because, as they increase, the animals breed less and die sooner: after all, we don't find ourselves disappearing under plagues of mice, rats or rabbits despite those animals' prodigious capacity for reproduction. At one point, scientists worried that climate change would act in the boar's favour and they'd be stampeding through our suburban gardens thanks to better breeding conditions. It's true that females can breed earlier and more often in those circumstances, with survivorship also increasing.[4] But when we study the long-term consequences of climate change in Central European populations, we find that these immediate gains translate into long-term losses. Younger and lighter sows have lightweight babies that grow up to be lightweight mothers with fewer and lightweight babies, dampening the rate of population growth.[5]

Your stereotypical pink porker doesn't especially resemble a wild boar. The family likeness can be glimpsed from certain angles – we're not talking chihuahua and wolf here – but there are also conspicuous differences. A boar has a narrow head and a long straight snout, pointy ears, and a tail that never corkscrews. The domestic pig, on the other hand, tends to have floppy ears, a corkscrew tail, and a podgy face with a short snout that looks like it's just run headlong at great speed into reinforced glass. The boar's weight is forward; the pig's is on its rump and flanks, befitting its function as a maker of meat. In these respects, the pig is a remarkably recent invention. As Richard Lutwyche points out, until the eighteenth century the British pig was a lacklustre variation on the wild boar, with a long snout and long legs;[6] not until the arrival of Asian pigs on new trade routes did the body become plump and the face squashed. Perhaps the best-known difference between pig and boar, though, is the dentistry. Both male and female boar boast a fine set of tusks, the female's lower and the male's upper and lower; the upper tusks are hollow and are used to sharpen the lower pair. Farmers aren't especially keen on keeping tusked pigs, which is why, as no less an authority than Charles Darwin noted, 'the boars of all domesticated breeds have much shorter tusks than wild boars'.

In a natural landscape without human intervention, none of these differences will stick. The boar is perfectly fitted to its environment; the same can't be said for pigs. Release pigs or hybrids into the wild and, assuming they survive at all, selection pressures will ensure a rapid reversion, after not very many generations, to boar-like behaviour and boar-like appearance. This experiment has been carried out countless times, often by accident. Schoolchildren obliged to fret over mankind's innate wickedness in *Lord of the Flies* would be better employed contemplating the vital question: namely, how did that colony of pigs end up on a small coral island? The answer – although the novel doesn't say – is that sailors would deposit pigs in remote locations so that they could rely on a ready source of meat the next time they were passing. Pigs have been scattered in the most unlikely of places and left to fend for themselves, usually at a terrible cost to local biodiversity: pigs introduced by sailors to Mauritius were largely responsible for the extinction of the dodo. Observing some of those porcine outposts, Darwin reported changes of colour, body shape and muzzle, as well as the lengthening of tusks. A few thousand years of domestic breeding are no match for the forces of evolution.

But about those tusks . . . exactly how dangerous are wild boar? The answer, in a word, is *not*. Wild boar are listed in the UK's Dangerous Wild Animals Act alongside such terrifying beasts as the tapir, aardvark and giant anteater. You can't be too careful;

nipping down the road for your newspaper one morning, you might get savaged by an aardvark. It's certainly possible that in Britain, sooner or later, probably *much* later, a boar will kill a human. In that same period, tens of thousands of people will die on the roads, hundreds will be trampled to death by cows, and dozens will meet their maker having fallen over while trying to put on their socks. (There are between five and ten such deaths each year in the UK, although admittedly we're using 'socks' generically to include stockings- and tights-related incidents.) Boar are considerably less dangerous than dogs. If you're really determined to be killed by a boar, you could increase your chances by hunting them; even then, you're far more likely to get shot by another hunter – the cause of up to twenty deaths each year in France alone – or have a fatal car accident while travelling to the site. Boar are at their most dangerous when crossing the road; they're regularly involved in collisions, in which of course they tend to come off worst. (That said, the only deadly incident involving wild boar in modern Britain occurred in 2015, when a driver on the M4 was killed after hitting a boar that had escaped from a nearby farm.) Significant urban boar populations exist in cities like Berlin and Barcelona, but while there are many reasons why such close proximity is far from ideal – including the car crashes – violent confrontations with humans are exceptionally rare. Boar are wise enough to keep out of our way. In the Forest of Dean, one of our friends came round a corner to see a boar twenty metres ahead, ambling straight towards him. He stopped, the boar stopped, and after a moment's pause each of them started edging slowly backwards.

There have been several cases in the Forest of people suffering minor injuries to their hands after contact with boar – normally a sign that they were feeding the boar at the time. More serious incidents tend to involve animals other than humans. Sows with piglets are particularly uneasy when dogs approach, and, if they can't retreat, they'll attack first and discuss the rules of engagement later. In one case, a dog was killed in the Forest after running into some undergrowth amidst much barking and cornering a boar, which, sufficiently provoked, came out of the thicket tusks-first. Lambs have been hunted and killed by a pack of boar in the Forest – or so several journalists have reported. Judging by occasional accounts from other countries it's not completely impossible, although in Britain eyewitnesses only describe seeing boar eating dead lambs rather than killing live ones. The overwhelming likelihood is that they were scavenging: if they find a miscarried lamb they won't let it go to waste. Boar are true omnivores and eat pretty much anything. As aficionados of gangster movies will know, the sure-fire way to make a body disappear is to feed it to the pigs.

Sad to confess, but the primary reason why boar are considered dangerous is that we've killed everything else. Try telling a sub-Saharan farmer that we have dangerous wild animals roaming around Britain, and see what response you get. The cave hyena, the cave lion, the grey wolf, the brown bear, the cave bear, the wolverine, the sabre-toothed cat – all were native at one time, and all were hunted to extinction. (Hollywood voiceover: 'Who are the real monsters here?') As a nation we dislike risk, no matter how infinitesimal or imaginary, far more than we love wildlife. Boar are perceived to be the most dangerous of all the species still standing, and so they've become the repository for exaggerated fears.

Literature and myth have hardly helped. When boar feature in stories, it's usually as a food source: Asterix's friend Obelix maintains his strength by hunting and feasting on boar with a voracity that would make Henry III blush. Boar are routinely portrayed as ferocious adversaries in order to emphasise a hunter's courage. Occasionally, they're even allowed to win. The Australian horror film *Razorback*, which we've watched so that you don't have to, features a massive boar crashing through the outback looking for people to kill and eat. *A Game of Thrones* describes a scene of mutually assured destruction when King Robert is mortally wounded by a boar as he kills it; his dying wish is that the boar be served at his funeral feast. The most famous boar-related fatality of all involves Adonis, the handsome youth who, Ovid tells us, would rather go hunting than hang around having sex with the goddess Venus. She begs him to hunt unthreatening animals like hares instead, and specifically warns him that boar have the power of a lightning bolt in their tusks, but of course he takes no notice and, in short order, gets himself gored to death. So far, so tragic, but when Shakespeare expands this story for what would become his first published poem, he turns the eroticism up to eleven and adds an undertone of sadomasochistic bestiality for good measure. Venus in her grief decides that the boar must have been trying to kiss Adonis: 'And nuzzling in his flank, the loving swine / Sheathed unaware the tusk in his soft groin'. Contrary to rumour, you're unlikely to come across those sorts of goings-on in the Forest of Dean.

Our guide through the Forest that unseasonably mild October day was David Slater, the wildlife photographer best known for his starring role in what is now immortalised as the 'Monkey selfie copyright dispute'. (Google it: you wouldn't believe us if we tried to explain.) David has been living in the area for over a decade, during which time he's become one of the strongest advocates for the continuing presence of boar in the Forest.

Wild boar are the awkward squad; they're useless at winning friends and influencing people, and the adults aren't blessed with movie-star looks. (Piglets, known as humbugs because their coloration resembles the boiled sweet, are much cuter ambassadors than their gnarly parents.) There's a Beaver Trust, a Lynx Trust, a Red Squirrel Survival Trust, an Otter Trust, a Badger Trust, a Wolf Conservation Trust, and so on; *nobody* trusts wild boar. Even rewilders treat them like an embarrassing uncle at the family reunion. Often we've read how beavers are the first successful mammal reintroduction into Britain in modern times, boar having apparently been airbrushed from the record. That's why boar desperately need champions like David Slater if they're to avoid the unprecedented fate of having been driven to extinction twice.

Taking us into the darkest depths of the Forest, David estimates that we have an even-money chance of spotting wild boar. We're happy with those odds: the Forest of Dean is forty-two square miles of woodland, and the boar could be anywhere. In fact, they've already been *everywhere*, because the grass verges are churned up for miles around. That may look like a promising sign but it doesn't greatly increase our prospects: a small number of animals can do an awful lot of digging. There are two other factors that count against us. Owing to hunting pressures, the boar have become increasingly shy and nocturnal; they spend around twelve hours every day sleeping under cover. There have been anecdotal accounts from Europe that Covid lockdowns have emboldened boar to go marauding around farmland in broad daylight, but nothing like that has been reported in the Forest. Our other difficulty is that they'll smell us unless the wind is working in our favour. Their olfactory senses are 2,000 times more powerful than ours. Boar can detect scents five miles away, and five metres underground. Pigs would replace search dogs if only they showed the slightest willingness to be trained. They've been used for truffle-hunting in France and Italy, but the drawback is that they tend not to stop until they've eaten the truffle. Worse than that, their human companions often lack a full complement of fingers, having tried at some point to prise the bounty from a pig's teeth.

Almost the first thing that happens as we set off through the Forest is that Tim spots a goshawk darting between the trees. With the same dubious logic that motivates wildlife lovers the world over, we conclude that this must be a good omen. Goshawks, like boar, were persecuted to extinction all over the British Isles, but they started to reappear in the 1960s either as escapees or deliberate releases. They're fractionally smaller than a buzzard and far less conspicuous: whereas buzzards ride high thermals as they look for scavenging opportunities, goshawks hunt fast and low, taking birds and small mammals such as voles, rabbits and squirrels. Further acquaintance may reveal a cuddly side, but, as even their greatest champion Helen Macdonald reports, at first they seem like 'merciless, psychopathic killers'.[7] The word 'goshawk' comes from the Old English *gōshafoc*, meaning 'goose-hawk' – not because our ancestors had drunk so much mead that they couldn't tell the difference, but because goshawks were trained by falconers to attack much larger species such as geese and cranes. Goshawks have done exceptionally well in the Forest of Dean and the neighbouring Wye Valley, with over sixty nesting sites across the region. It's a thrill to see one: boar or no boar, already the day has been a glorious success.

With all the talk of boar and goshawks, you'd be forgiven for assuming that the Forest of Dean must be a biodiversity hotspot. Thanks to Gloucestershire Wildlife Trust, Heritage Lottery funding and a team of volunteers, it's certainly one of the best-surveyed areas of Britain. You name it, they've counted it. We know where the veteran trees are (there are depressingly few); we know which butterfly species are present (several flutter on the edge of local extinction); we know that reptiles such as adders are in trouble. Huge effort is devoted to documenting an already depleted bioscape. Compare the Forest of Dean with the bared hillsides of Dartmoor or the Lake District and you'll be grateful for its richness; glance across to mainland Europe and you'll appreciate why our forests should be a cause of deep national shame. Take, for example, one of Croatia's eight national parks: Plitvice, a UNESCO World Heritage Site founded in 1949. It's nearly three times the size of the Forest of

Dean, but out of all proportion healthier and more vibrant and abundant: it boasts wolves, bears, lynx, wildcats, and, yes, beavers and boar. By every measure, Plitvice has more biodiversity than the whole of the UK put together. It's not our job to do the work of the Croatian Tourist Board, but these contrasts expose any deception that our own forests are wildlife havens. As for financial costs and benefits, Plitvice employs a thousand people, and its 1.5 million annual visitors pay up to forty euros each, per day, towards the maintenance and improvement of the park. With an ambitious vision, you can save your corner of the planet *and* make a living. Without it, you can eke out a profit from commercial logging, antagonise the locals, and shoot wild boar on the side to cover the wages of a couple of rangers.

If there's no British Plitvice, it's because we seem incapable of thinking about our royal forests and national parks in that way. For too long, the relevant agencies' idea of reforesting was to establish non-native conifer plantations on peat bogs, as if trying to demonstrate that you can plant trees and still destroy the environment in the process. Any government wanting to leave a positive legacy could simply instruct the Forestry Commission to repurpose our forests for the prioritising of biodiversity: it's that easy. The good news is that ecologists' voices are finally being heard, with forestry agencies providing some of the strongest support for initiatives such as pine marten reintroductions (about which more in the next chapter). Yet every victory has to be won against a default policy of intensive monoculture timber production. We mustn't forget that what little forest survives in public hands has never been maintained for wildlife, unless of course it can be shot. We may feel safe in the knowledge that we're not going to get eaten by the Big Bad Wolf, but that's part of the problem. Far from being natural spaces, our forests are heavily and often ruthlessly managed, and for nigh on a thousand years we've been entering them only under sufferance. No wonder we've forgotten what a native woodland looks like: the forests rated most highly for what the jargon calls 'ecosystem services' comprise non-native conifers close to urban areas, complete with cycle paths, plenty of dog-walking opportunities and a nice coffee shop.[8]

After they'd finished slaughtering Saxons, almost the first thing the Normans did was to introduce Royal Forests and Forest Law. The word 'forest' specifically meant an area of land (usually, but not necessarily, woodland) set aside as a preserve for the king's hunting. This was a land-grab, pure and simple, and the peasants unfortunate enough to live within forest boundaries laboured under brutal rules. Kill a wild boar and you could be punished with blinding. In case that wasn't a sufficient disincentive, by the end of the twelfth century castration was enforceable, too. You were allowed to keep a mastiff as a guard dog, but only if you cut off the balls of the forefeet so that it wouldn't go chasing after deer or boar. The boar, then, encapsulated debates about entitlement, property and rights. For the aristos, it represented courage and ferocity as a noble adversary in the hunt; the Earls of Oxford even took to insisting that their family name, de Vere, came from the Latin *verres*, meaning 'swine'. Richard of York went into battle at Bosworth under the banner of a white boar, later inspiring Shakespeare to describe him as a 'wretched, bloody and usurping boar'. Shakespeare was tapping into the grievances of the dispossessed, for whom the boar represented destruction, oppression and misrule, a living reminder of their powerlessness in the face of irresistible forces.

Throughout the centuries, wealthy landowners have reintroduced boar into forests and hunting estates, and the peasants have borne the brunt of any resulting mayhem. Those swinish de Veres kept boar in Chalkney Wood, Essex, as a consequence of which 'the Inhabitaunts thereabouts sustained by them very greate losse and damage'. James I released them into Windsor Park, and instructed that the land should remain unploughed so that his horse wouldn't stumble while he hunted them. His son Charles set them loose in the New Forest, where their population grew tremendously; the Civil War gave locals an opportunity to wipe them out, but not before the boar had (in John Aubrey's words) 'tainted all the breeds of the pigges of the neighbouring partes, which are of their colour; a kind of soote colour'. Aubrey himself tried keeping boar, with mixed success: 'they digged the earth so up, and did such spoyle, that the country would not endure it: but they made incomparable bacon'. A century later, the parson-naturalist Gilbert White reported another ill-fated reintroduction. Boar imported from Germany had been turned loose in woodland 'to the great terror of the neighbourhood', until locals took matters into their own hands: 'the country rose upon them and destroyed them'.

So on whose authority are there boar in the Forest of Dean today? It's worth bearing in mind that boar are officially classified as dangerous, and individuals have been prosecuted for releasing them. Say what you like about the Forestry Commission

– and people often do – but it's admirably upfront about the extent of its rights and the limit of its responsibilities:

> A common complaint from landowners, including private homeowners, to the Forestry Commission is that feral wild boar are getting from the Forest onto their land. However, all property owners need to note that maintenance of their boundaries in a boar proof condition is their responsibility.[9]

What this amounts to is: don't bother trying to sue us. The Commission sells places on guided hunting weekends, and it generates a tidy income when the meat makes its way onto plates in chi-chi London restaurants, but any collateral damage is none of its business. And so the latest instalment in the centuries-old saga of high-handed elites and long-suffering locals plays itself out. All the more credit, then, to the majority of residents in the Forest of Dean who remain sanguine about boar: in surveys, two-thirds of them claim to be unconcerned by their presence, and more than three-quarters are 'excited' or 'interested' when they see a boar.[10] A public consultation in 2006 found more than half of respondents insisting that – in Defra's own gritted-teeth paraphrase – 'wild boar are a former native species [and] have a right to exist in the countryside'. Even allowing for differences in wording, the survey numbers compare favourably with the mere thirty-eight percent of the general population who support the reintroduction of wild boar into Britain. Perhaps, after all, to know boar is to love them.

*

We abhor the very suggestion that boar are boorish or boring. Here are some facts that prove otherwise:

Scrofula, a disease in humans better known as the King's Evil, has nothing to do with wild boar despite being derived from their scientific name, *Sus scrofa*. Etymologists disagree about the reasons. Either it's because the swellings in the neck look like little pigs, or because pigs were believed to suffer from the disease. This feels like a choice between two wrong answers.

A variant of the Saxon word *eofor*, meaning wild boar, survives in Dutch as *ever* or *everzwijn*, and lurks just below the surface of a number of English place names: Everleigh, Eversholt, Everton. Between the sixth and the ninth century, York was Eoforwic: 'wild boar settlement'.

Eat a boar's testicles if you want 'to invigorate the body, and to help the work of generation'. Coughing blood? Dry and drink some boar dung in a tasty 'liquor' and you'll be cured in no time. Rub some specially prepared boar urine on the navels and noses of young children if you want to rid them of worms. While you're at it, try a sip of that urine yourself and you'll have no more problems with gallstones. Don't just take our word for it: the early-eighteenth-century scientist Richard Bradley says as much.

Boar have a special place in the Japanese national psyche. Hachinohe's Wild Boar Famine of 1749 saw an estimated 3,000 people die of hunger. The problem had been caused by farmers switching to soybean crops and abandoning land when yields dropped. The boar tidied up the remains of this exciting new delicacy and, as their populations soared, they started on the other crops and invaded the town looking for more food. Samurai led campaigns against them, and the chaos continued for over a decade. Apparently there's a traditional saying in Japan: 'When you get married, choose a place where there are no wild boar.'

The French and Italian words for wild boar are, respectively, *sanglier* and *cinghiale*. Both come from the Latin *singularis*, as in *singularis porcus* (singular or solitary pig). This is because the male boar keeps to himself, and only starts being sociable when he decides it's time to mate. Sows, on the other hand, prefer company: two or three will band together with large numbers of piglets in groups called sounders.

Male boar have a peculiar way of demonstrating to females that they're hungry for love: they froth at the mouth. Humans have mistaken this for a sign of aggression, but in fact it's to release pheromones that sows find irresistible. When the male is ready, he'll nuzzle the sow's flanks and rump, and grunt rhythmically in what has been kindly described as a mating song. Boar, like dogs, lock together during sex, and stay that way until the male has calmed down a bit.

If boar can't run away, they work on the principle that the best form of defence is attack. Even so, they've no interest in picking a fight. They're famous for their mock-charges, and it's a very foolish human who waits to find out whether or not they mean business.

Boar are surprisingly strong swimmers. There are records of them swimming twenty miles or more. Perhaps, after all, those pigs in *Lord of the Flies* made their own way to the coral island. Plenty of people dream about swimming with dolphins, but if you'd rather swim with pigs you can visit Pig Beach, an uninhabited island in the Bahamas that's been home to a colony of wild pigs for hundreds of years. They enjoy swimming, no doubt partly because it helps them to cool down, and they'll gladly piggy-paddle some distance offshore to tourist boats in the hope of getting rewarded with food.

You have to look long and hard to find scientists celebrating the engineering prowess of boar. Read the academic literature and you'll conclude that they're nothing more than disease-addled crop-wreckers. An EU-funded network, the italically challenged E*NET*WILD, dresses its website with cute video of boar rootling, but its stated aim is to produce robust data on the distribution of the species so that the risk of disease outbreak can be controlled. Boar are interesting to this particular brand of scientists because, as the website explains, they're a 'host species' for 'pathogens', so we need to know when and where to wipe them out. In the wake of Covid – when paranoia about zoonotic diseases has become particularly frenzied – this kind of research is lucrative. Careers are made, institutes are founded, and there's always a wine reception at the international conferences. Meanwhile, scientific studies of the role of boar in natural ecosystems barely exist.

So, as the song nearly said, let's hear it for the boar. Like beavers, they're amazing. One of the boar's most eloquent supporters, George Monbiot, has argued that they're 'the untidiest animals to have lived in this country since the Ice Age'.[11] Untidy is exactly what the environment needs to become. No other British species creates the ground disturbance that's a natural part of woodland ecosystems. With their large heads and stocky, powerful build, boar are the bulldozers of the animal world. They're supremely adapted for rootling. Thanks to a specially elongated shelf of bone in the skull, the snout can be moved independently of the chewing muscles, making it easier to snuffle or truffle. That characteristic piggy nose is perfectly designed for ramming through earth, and the length of the snout makes it an ideal lever. The cartilaginous disc at the end is supported by a small pre-nasal bone found only in the pig family, and closable nostrils protect boar from the occupational hazard of getting soil up the nose.

Despite an undeserved reputation for being fat and lazy, boar and pigs have to work hard at feeding. It takes an awful lot of earthworms and beetles to fill them up, and only rarely do they stumble across Mafia victims. Around ninety percent of the boar's diet is plant matter.[12] Their problem is that they have just a single stomach compartment, whereas most large herbivores (deer, cattle, sheep, giraffes, camels) have complicated multi-chambered digestive systems that are designed to squeeze every last drop of goodness out of a leaf. This sets the boar on a quest for energy-rich foods, which is why they take such delight in chomping their way through crops such as potatoes and maize. In spring and summer, when food availability is less precarious and grasses are at their most nutritious, boar tend to graze. They rootle much more during the colder months when grasses and other leafy plants have little nutritional value. Acorns and beech seeds are particular favourites, as are roots and tubers where plants store their energy in winter.

The grazing done by boar is a natural part of our ecology, but we have to admit that there are plenty of other animals, from rabbits to red deer to sheep, already grazing (or more likely overgrazing) our landscapes. We don't need another grazer. Rootling, on the other hand, is uniquely and unmistakably boarish, and it's vital for restoring ecosystems. As Lake District sheep farmer and author James Rebanks advises, 'if you have woodland, get pigs in it';[13] you'll be rewarded with an explosion of botanical richness. Badgers occasionally turn over grassland, and are especially partial to newly laid turf, but they're mere lightweights. Boar are the only species capable of the heavy lifting. This is crucial because the British Isles is, by an impressive margin, the most deforested region of Europe. Our politicians love to be seen planting trees, but before that photo-op can take place, someone or something has to prepare the ground. The same holds for the overwhelmingly more effective – if less photogenic – method of reforesting: allowing the natural regeneration of native species from fallen tree seed. The trails created by wild boar, which look like the work of someone drunk in charge of a small digger, provide ideal nursery beds. To finish off the job, their ridiculously small stiletto-feet are the perfect tools for pushing seeds down into the soil where they can germinate. Even dense species such as bracken are no obstacle to a boar on a mission, because starchy underground rhizomes are a favourite food, and boar will rootle deeply to get at them. For us humans, it's – to say the least – character-building to break up the tangled rhizomes and clear the litter layer so that other species become established. It can require repeated cutting, clearance, and treatment with herbicides over five years to break through the dense mat of bracken. Boar do the work quickly, and for once they're happy to oblige.

An experiment by Fiona's colleague Chris Sandom in the Alladale Wilderness Reserve of Scotland has shown just how effective boar are at creating patches suitable for natural seed germination in the toughest of habitats: dense heather moorland.[14] Each boar uprooted an average of forty square metres of heather per week. In this case, the boar were fenced in and given supplementary feed. Nevertheless, the costs were comparable to other methods of clearing the ground, and, of course, *wild* wild boar wouldn't need fencing or feeding. An increase in the density of boar increased the area disturbed rather than the intensity of the disturbance, which matched our observations in the Forest of Dean: only a few areas of pasture were completely broken up, but there was widespread disturbance across all the areas we visited. Boar don't have an especially large home range, but they cover much of it daily, presumably because resources such as acorns, beech mast, chestnuts and grubs are commonly distributed across woodlands. It wouldn't make sense to keep digging in the same place.

We ought to mention bluebells, for which the anti-boar lobby has developed a convenient passion. Britain is renowned for its bluebells, and in spring you can't pick up a country magazine without encountering a soft-focus photo of them. These monocultures aren't natural; they're a sign of our damaged and diminished ecosystems. The evidence from the Forest of Dean is that bluebells coexist with wild boar, but more patchily and not as uninterrupted carpets that smother and suppress less vigorous species. In other countries where boar are found in far larger numbers than here, woodlands enjoy greater plant diversity, owing in part to the opportunities provided by the complex habitats created by boar. Soil disturbance turns up seeds buried in lower layers so that they can germinate. We can't put it better than Forestry England (a subsidiary of the Forestry Commission), which freely acknowledges on its website that, at the right population density, boar will have a 'positive impact' on forest ecology:

> Ecologically, boar at low densities are good for the natural environment. The rooting and wallowing behaviours break up static ecosystems and allow an increased range of plant species and insect fauna to grow.[15]

Boar are also fantastic seed-dispersers. Acorns and beech mast rarely emerge from a boar's digestive tract unscathed, but many plant species do survive that torrid journey. Other seeds hitch a ride in the boar's bristly fur, a mode of dispersal for which the sadly-too-long-for-Scrabble word is epizoochory.

Not everyone's going to be a winner when boar are in the neighbourhood. A torpid slow-worm, the eggs of ground-nesting birds, or a hibernating dormouse may be easy pickings. At the moment, good studies of the complex interactions between boar and other vertebrates are woefully scarce – and urgently needed. Boar may be eating reptiles in the Forest of Dean, and we know for a fact that they're eating frogs. While it may be scant consolation for the individuals being eaten, the credits seem to outweigh the debits for their species generally. The restoration of long-lost habitat complexity through the bioturbation wrought by boar is generally beneficial to wildlife. The clearing of dense vegetation creates areas for reptiles to bask, and abandoned wallows form small ponds valuable for frogs, toads, newts and any number of aquatic invertebrates.

One of the British landscape's success stories of recent times is intricately linked to porcine landscape engineering. The Knepp estate in West Sussex is rightly lauded as exemplifying what can be achieved for biodiversity through a rewilding of lowland Britain. Blocked from introducing boar because of the Dangerous Wild Animals Act, the owners released Tamworth pigs to fulfil the same ecological function. Knepp now has the largest breeding population of purple emperor butterflies in the country, from a starting point of zero, and it's all thanks to pigs that they're thriving. One of the most spectacular butterflies in Britain, the purple emperor looks more like a tropical species than something you'd expect to meet in the south-east of England. It feeds on aphid honeydew and tree sap rather than nectar from flowers, and comes down to extract salts from animal dung or road surfaces. In summer, the males put on displays for females above tree canopies, and aggressively chase away intruders. They'll even attack goshawks and buzzards in a deadly game of bluff. Purple emperor butterflies are prospering at Knepp owing to the availability of sallow scrub, which the females need for egg-laying; the bare earth produced by the pigs turns out to be ideal habitat for sallow to colonise.

The authors of the periodic reports on biodiversity in the Forest of Dean lament the scarcity of habitats suitable for the small pearl-bordered fritillary and the dingy skipper butterfly; both species are subject, locally and nationally, to long-term decline. The reports note that overgrowth with bracken and (for the fritillary) the drying out of habitats are contributory factors. Butterfly Conservation recommends that the dingy skipper can be saved by a two-pronged approach: turf-stripping in grasslands, and creating, in woodland rides, patches of bare earth that can be colonised by food plants. If only there were a native species capable of clearing the bracken, exposing the earth, and making wallows where rainwater can collect . . . Wait a minute! The Dutch have already carried out research to prove that the (closely-related) grizzled skipper benefits from the microhabitats created by boar.[16] So now that boar have finally spread across the Forest of Dean, we should just let them get on with doing what they do best. Something different is happening all across mainland Europe, where – as we're about to see – people are busily terrifying them, scrambling their breeding patterns, erecting fences, and introducing disease into their populations.

*

The truce in the Forest of Dean shouldn't imply any governmental enthusiasm for boar. Defra's vets are instinctively a risk-averse breed, and their – let's be tactful – *unscientific* support for the badger cull as a way of reducing bovine tuberculosis demonstrates a bloodthirsty readiness to protect livestock at the expense of native species. As far as Defra is concerned, whatever can't be controlled is a menace; boar are just a potential reservoir for a number of zoonotic diseases that would cause havoc if spread to farm animals. For example, in several European countries boar carry bovine tuberculosis. There are no reported cases among boar in Britain, but given the threat to the cattle industry and the £50 million that TB control costs the British taxpayer every year, they've been put on notice. As the online TB hub endorsed by Defra makes clear:

> feral wild boar in England should be regarded as potential amplifier hosts for *M. bovis* [bovine tuberculosis]. If evidence emerges that feral wild boar are playing a significant role in the epidemiology of TB in England, measures could be taken to control the population.[17]

Note the emphatic repetition of 'feral': as Defra already insists on killing badgers in the face of concerted opposition from scientists and the general public, what hope is there for a less popular species that it refuses to acknowledge as native?

An even bigger worry is swine fever, or, rather, *fevers*. As if classical swine fever weren't trouble enough, African swine fever emerged in Kenya in 1910. The two illnesses have symptoms in common, but they're caused by different viruses (the classical being a small RNA virus, and the African a large DNA virus). Happily, non-porcine mammals, humans included, aren't susceptible to either, but for pigs and boar they're almost certainly a death sentence, whether from the disease itself or from the culling regimes that are integral to the control strategy.

Classical swine fever is highly contagious, via saliva, nasal droplets, faeces and urine. Just about any material that's been in contact with a virus-shedding animal can infect new pigs: bedding, pens, vehicle wheels, trailers and clothing will all do the job. African swine fever seems less easily excreted, but it's abundant in blood, which means that tick bites are an effective transmission route. In fact, tick–hog infection cycles are the natural way of things. Common warthogs, the sub-Saharan cousins of wild boar, have been rootling away in the midst of African swine fever for hundreds and possibly thousands of years. The virus usually causes them no particular issues, and piglets develop lifelong immunity after being infected at an early age by ticks in their burrows. Humans, inevitably, created the problem: when farmers began to raise domestic pigs in areas of warthog habitat, the ticks tucked into this plentiful new food supply, and promptly infected the pigs with African swine fever.

With the exception of the Iberian Peninsula, ticks are rarely a source of infection for domestic pigs across Europe. Infected pork is the main culprit for both swine fevers. The first European outbreak of the African variety, in 1957, was probably caused by feeding pigs with swill from ships in international harbours.

Although giving pork products to domestic pigs and wild boar is banned by the European Commission, in practice it's impossible to police supplementary feeding and baiting by hunters – or even by wildlife lovers who want to get close to a friendly boar. At several picnic sites in the Forest of Dean, the boar are noticeably bolder because, contrary to all the official advice, people go there to feed them.

Currently, there's no safe and effective vaccine for African swine fever, but for classical swine fever a vaccine can be administered to pigs (via injection) and wild boar populations (via baited food). In Britain, vaccination isn't part of the government strategy to prevent classical swine fever: the official report from Defra and the Scottish and Welsh governments states with swivel-eyed logic that 'a vaccination campaign distracts vets from the primary task of identifying and controlling swine fever'.[18] Their real concern is with the difficulty of distinguishing vaccinated from infected animals, which might have consequences for export markets. Vaccination is therefore a last resort – a firebreak to check an epidemic rather than a policy to be deployed prophylactically. At the very least, risk-averse Defra is taking a massive gamble. We already know that, in the wild boar populations of Eastern Europe, survivors are likely to become long-term carriers. Not all wild or domestic pigs with classical swine fever display symptoms, and those animals can shed the virus for months before the disease is ever detected.

An outbreak of classical swine fever in the Netherlands in 1997 resulted in the culling and incineration of around 700,000 infected pigs and the proactive slaughter of a further 1.1 million animals. (Another seven million were killed locally for animal welfare reasons connected to the disruption of markets.) Thankfully, classical swine fever hasn't been detected in domestic European pigs since 2014. African swine fever now offers an altogether more severe level of threat. It was eradicated from mainland Europe in 1995 but reappeared in Georgia in 2007, and has become widespread in East Asia, where an outbreak in the summer of 2018 saw China's pig herd halved. That's a ballpark figure of 300 million dead pigs, or more than one-fifth of the world's pig population. Regular outbreaks of ASF are now being reported among wild boar in Eastern Europe, and also in Belgium and the east of Germany. As far as domestic pigs are concerned, most European incidents have taken place outside the EU, affecting small numbers of animals in backyard holdings. A notable exception involved a single large commercial farm in western Poland with 23,700 pigs – a number that reveals as much about the cruelty of intensive farming as it does about disease epidemiology. Large-scale food producers, and the politicians they lobby, aren't ecstatic about either of the swine fevers: they can wipe out an entire industry in a matter of months, with desperate consequences for jobs, the economy and animal welfare.

If you wanted to design the ideal conditions for African swine fever to spread, all you'd need to do is copy hunting practices across Europe. They're a catastrophe not just waiting to happen, but already happening. Hunters protect wild sows and establish winter feeding areas, so that boar populations become artificially abundant well beyond the environment's natural carrying capacity. So much for any claims that we should hunt boar to keep the numbers down; in parts of Europe, boar numbers are rising quickly and unnaturally because of hunters who see them as tasty vermin. The feeding of boar and the deliberate planting of crops adjacent to woodland have put a stop to seasonal population cycles. Boar are now roaming the outskirts of Barcelona in search of discarded fast food, because they're dispersing from the hills beyond the city where their numbers have grown out of control owing to supplementary feeding. No one seems inclined to learn the lessons of the Japanese wild boar famine.

Hunters typically kill and eviscerate animals in the field, sling the carcasses into the back of a truck, and leave behind blood and what is decorously known as offal. In the process they contaminate the soil, hunting premises, clothes, vehicle tyres, their own boots, knives and other kit. Once it's in a landscape, African swine fever persists for an awfully long time. It can survive for fifteen weeks at room temperature, many months at 4°C, and indefinitely when frozen. Camera-trap surveillance tells us that boar are touchingly sentimental when it comes to eating Uncle Wilbur or Cousin Peppa: they nudge and nose but don't nosh the carcasses. Unfortunately, even that level of contact with a diseased animal may be enough. They'll certainly gobble down any offal left by hunters, and rootle in blood-contaminated soil.

One obvious way to prevent the spread of disease is to reduce population density. Fewer animals in the landscape means that fewer come into contact with an infected individual. In Poland, the initial response to African swine fever was to encourage hunters to go in all guns blazing. This was a seriously bad plan for several reasons. The disruption caused by hunters resulted in animals fleeing their normal territories, taking the disease with them. (Something similar has happened in Britain with the badger cull and its so-called 'perturbation effect'.) Where older sows get taken out, younger animals come into breeding condition more quickly. What's more, reducing the population density by hunting results in better survival of new piglets: compensatory reproductive success is a phenomenon familiar to any first-year ecology student, even if not always to governments. These piglets are susceptible to the virus and fuel the epidemic. Then the hunters give the virus the opportunity to hitch a ride on vehicles, boots and clothes. The Polish authorities report that African swine fever has, as they put it, 'jumped a considerable distance' within the country on two separate occasions.[19] Well, who could have been responsible for that? What was initially a local outbreak has now spread to over one-third of Poland. Meanwhile, Belgium has solved its swine fever problem with a strategy that included the opposite approach: it banned recreational hunting in affected regions.

Governments are taking swine fever so seriously that they're throwing up fences across the once proudly borderless EU. In 2019, Denmark completed the construction of a 1.5-metre-tall, 0.5-metre-deep fence that runs along the entirety of the Danish–German border. The €4 million bill was considered money well spent by a government determined to protect its 5,000 pig farms and exports of around 28 million pigs a year. Even so, boar still have the option of road and rail crossings, and if they fancy a dip, they can (and occasionally do) swim across the Flensborg Fjord, which separates the two countries. Suitably inspired, Germany has set about building its own fence to stop the incursions of ASF-infected boar from Poland. The problem is that fences can have calamitous ecological impacts because they divide populations and sever migration routes.

Perhaps, after all, the European disaster shows us that the Forestry Commission is the least worst of all the organisations that boar could encounter. When boar turned up on National Trust land at Stourhead in Wiltshire, the Trust (visit their website to 'Find out how we're helping to look after natural habitats and native species') hired marksmen to exterminate them forthwith, camouflaging the news with dollops of euphemism in case their more squeamish members couldn't bear very much reality: 'we have taken the difficult decision to remove the animals from the estate.' The Trust's behaviour hammers home the vicious consequences of that word 'feral'. Because boar have no legal protection, landowners are entitled to shoot them. There's no closed season: pregnant sows, nursing sows, piglets – they're all, quite literally, fair game. In 2008, Defra did manage to cobble together a Feral Wild Boar Action Plan that might have taken a view on all of this, but never before has a document so abjectly failed to deliver what it says on the tin: the only 'action' was to turn inaction – 'no government intervention' – into official policy.

As part of his doctoral research, the geographer Kieran O'Mahony went to the Forest to speak to residents, government agencies, local councils, farmers and conservationists about the boar. On both sides of the argument, residents clearly identified the term 'feral' as a weapon, delegitimising the status of boar as wild and native animals: 'they aren't meant to be there . . . They are feral because . . . they have gone wild rather than being truly wild ones'; 'they have been here long enough now . . . many successive generations born in the wild . . . the feral label should be taken away. Now they are "wild" wild boar'.[20] The government's own allegiances are plain enough: the misnamed Action Plan talks about boar and 'native biodiversity' as two different entities, while Schedule 9 of the Wildlife and Countryside Act splits wildlife into three categories: 'native', 'non-native' and 'animals no longer normally present'. This enigmatic final category, unrecognised by international law or any of the conventions to which we're signatories, contains boar and beavers. Deliberately releasing these animals into the wild, or simply failing to stop them escaping from captivity, has been made an offence.

The miracle amidst all these official prejudices and manoeuvrings is that wild boar are still in the Forest. The Forestry Commission has got any number of things wrong – the culling strategy, the public relations, the lack of a long-term plan – but *boar are still in the Forest.* It's worth repeating because it defies explanation. How easy it would have been to behave like the National Trust, 'removing' the animals from the land and saving itself an ongoing headache. It's a fair assumption that the Forestry Commission has resisted immense pressure not only from external forces but from its paymasters in Defra. There are boar in the Forest of Dean and – it seems – heroes in the Forestry Commission, even if we don't know who they are.

We'd been walking for nearly six hours, and the light was fading fast. David had taken us to some of the most likely locations, where we'd found wallows, recently uprooted turf, scats (otherwise known as poo), and tracks left by an elusive species with a rounded hoof and a characteristic dewclaw. Grey squirrels were everywhere, and we'd spotted roe and fallow deer. (If the authors of this book ever get divorced, their irreconcilable difference will be whether fallow deer should be considered native – a debate that we'd better not rehearse right now.) What we hadn't found was actual boar. Our daughter had peeled off back to the hotel, and David had a home to go to as well. After goodbyes, we decided to walk one more time down a track so straight that we couldn't get lost even in the dark. Fiona was shining her red-light torch past the first row of trees when we spotted an animal's eyes low to the ground. The excitement was momentary: it was only another fallow deer. So we turned back, disappointed and amused, and, just as we were off our guard, we heard, no more than a couple of metres away in the undergrowth, a sudden deep primeval grunt. In that instant, we wish we could tell you that our first step was forward. Our *second* step was forward, but too late: the boar had turned tail and fled into the forest. It wasn't going to risk crossing paths with the deadliest predator of them all.

THREE

On the Trail of the Lonesome Pine Marten

Let's begin with the story of a single solitary pine marten, today running free through the forests of New Jersey. This marten is the Columbus of her species, having survived not just one or two but five perilous pioneering expeditions. It's worth recounting the series of extraordinary circumstances that brought her here after an idyllic youth spent in a Scottish woodland, because through her history we can begin to judge the prospects for the entire species as it makes its tentative comeback from near annihilation.

Pine martens are the beneficiaries of what may be the most carefully managed recovery programme in the history of British conservation. It didn't happen a moment too soon. There were around 150,000 pine martens in Britain during the Mesolithic period, making them the second most abundant carnivore after weasels. By the end of the nineteenth century, they'd all but disappeared, and in 1915 just three refuges remained: the far north-west of Scotland, the Lake District and a small patch in north-west Wales. Britain at that time, and for many decades afterwards, was a hostile and denuded landscape, with woodland cover at about five percent. Put simply, pine martens were running out of places to live, and were heavily persecuted across their shrinking range.

The 1995 population review of British mammals reported a total of just 3,650 individuals, the vast majority of them in Scotland, with England and Wales providing only double-digit contributions. Today that estimate is still probably about right, having dipped and then recovered in the interim, but for once there's more good news than bad. Granted much-needed legal protection in 1981, pine martens have also benefited from new afforestation strategies: woodland made up ten percent of England, fifteen percent of Wales and nineteen percent of Scotland in 2021, and although the increases have mostly comprised less-than-ideal non-native species and young trees lacking suitable holes for denning, they're better than nothing. As the pine marten's range north of the border has expanded, Scottish martens have been trapped and then translocated first to sites in Wales and now England. Outside Scotland they're still classified as Critically Endangered, but at last they've got a fighting chance of re-establishing viable populations in areas of mainland Britain where they'd been absent for decades, if not centuries.

Our friend Columbus took part in the second wave of the translocation project. She was caught and delivered to Mid Wales, fitted with a radio collar and released into the wilds. That initial journey seems to have instilled a wanderlust in her. Free once more and having weighed up her next move, she headed to the coast, seventy miles south-west. It's assumed that somehow she stowed herself away aboard a container ship, hunted mice and rats during a long Atlantic crossing, and didn't emerge until the ship docked in Savannah, Georgia. She was found running around the port, collar still attached, and deposited in a local zoo. The weather wasn't to Columbus's liking, so eventually she was sent north to cooler climes, where for several years she pondered her ignominious captivity while humans strolled past in search of the elephants and tigers. A lesser marten would have resigned herself to the vagaries of fate, but not Columbus. One day she escaped, and she hasn't been seen since.

Not all martens are so wayward. The majority can be trusted to behave more responsibly by expanding out from their release sites into new territories on *this* side of the Atlantic, where they spread the pine marten love and raise kits. Fifty of Columbus's Scottish peers have been relocated to Wales without feeling the urge to board boats and sail away into different time zones. We had a particularly pressing reason for tracking them down. With a single exception, Fiona has seen every native British land mammal in the wild. You name them – mice and moles and voles, shrews of all stripes, badgers and otters, weasels and stoats, Bechstein's and barbastelles, roe and red – she's found them all, if not here then abroad. (Tim's lagging behind, with three species still to tick off.) We've even seen the greater mouse-eared bat, which at last count had a total British population of one individual; we caught several of his Polish counterparts a decade ago while harp-trapping near Krakow. What neither of us had ever encountered was a wild pine marten. Laura Kubasiewicz, one of Fiona's collaborators on the 2018 population review, spent three years carrying out fieldwork in Scotland for her doctoral research on pine martens; in all that time, she saw a wild marten once. And if you're reading this and thinking, 'I see pine martens all the time,' or 'One ran across the road just in front of me last week,' we officially dislike you.

Our heads told us we had no chance, but our hearts were singing with optimism. Together with our long-suffering younger daughter, we were driving up to Mid Wales one wet October day, and come hell or high water we were going to see a pine marten.

Talking of hell, our precise destination was the village of Devil's Bridge in Ceredigion. The name comes from the legend – common across Europe in various guises – of a woman whose cow escapes to the other side of a steep gorge. The devil offers to build her a bridge if he can have the first soul that crosses. The woman throws some bread over the bridge so that her dog runs after it, with all the expected consequences. The Welsh name is Pontarfynach, meaning 'the bridge on the Mynach'. At least with that version we don't need to feel outraged on behalf of the dog.

Devil's Bridge is actually three bridges, stacked with a haphazard beauty on top of each other, dating from the twelfth, eighteenth and early twentieth centuries respectively. The structures span a deep ravine, at the bottom of which flows the Mynach. There are tiered waterfalls, with five major drops adding up to a ninety-metre descent. It's been a sightseeing attraction for centuries. Turner came to sketch it, and Wordsworth wrote a godawful sonnet full of 'incessant shocks' and 'throbbing rocks'. Just as enticing as all this natural glory is the presence of pine martens. Devil's Bridge is ground zero for the reintroduction of the species in Wales, because it was around here in the autumn of 2015 that the first batch of Scottish martens was released.

That auspicious event is marked in various ways locally. At the entrance to a nearby car park stands a memorial sculpture to Rob Strachan – 'an inspirational naturalist and conservationist' – whom Fiona worked with for several years at the Wildlife Conservation Research Unit in Oxford. It's a wooden trunk-like structure, atop which crouches an adult pine marten; further down, peeping out of a convenient hole, are two kits. A mile or so away, the railway station marks the eastern terminus of a line served by steam train to and from Aberystwyth. The old weighbridge office has been converted into a rather fine 'pine marten den', with information for visitors about the species and the recovery project. As for our own prospects of spotting a marten, the writing was literally on the wall:

> Today, in the woodlands of the Rheidol Valley and beyond, there is just a chance you may see a pine marten. It is a slim chance: they are elusive, tree-dwelling and largely nocturnal, but they are here – back from the brink.

As Muhammad Ali once said of his next opponent, he had two chances – slim and none – and Slim just left town. Then again, Ali lost that fight.

Our trip to Wales wasn't just a jolly. Fiona needed to collect scats for a research project that involved screening wildlife for Covid. We'd brought a crack team of experts to help us: Henry Schofield and Patrick Wright both work on the reintroduction project; Bridgit Schofield is a hugely experienced ecologist. Then there was our younger daughter, with the natural advantage of sharp eyes undimmed by age. If there were pine martens to be seen, perhaps we'd defy the odds. At the very least, we'd find their scats. When you're dealing with the Scarlet Pimpernel of mammals, even a smear of poo can seem like a win.

Pine martens flicker on the edge of public consciousness. Some people have never heard of them, and others assume that they're birds on the basis that they must be related to house martins. Even those of us who know that they're mammals, and follow the clue in their name to deduce that they live in woodland, may have little idea what these baby-faced assassins look like.

First, the family resemblances. Pine martens are mustelids – a name derived from the Latin word for weasel, *mustela*. It's quite an honour for weasels to have been recognised in this way, because they're the junior partner in the mustelid enterprise. (Our native British weasel is the junior partner even among weasels, being smaller than any other weasel species worldwide, which is why its full and rather insulting name is the least weasel, and perhaps also why it's so *very* fierce.) There are seventy-ish species of mustelid. It's hard to generalise about such a diverse group of carnivores, but they tend to have in common a long body and relatively short legs. The smaller ones are slinky, and anybody who has seen a badger running will know that it looks like a rug being pulled on an invisible string. The six mustelid species native to Britain, in ascending order of size, are weasel, stoat, polecat, pine marten, otter and badger; the invasive American mink is about the same size as a polecat, while the wolverine, which would be the biggest of the lot, became extinct in Britain about 6,000 years ago. Ferrets are domesticated polecats; although they're sometimes found in the wild as escapees, they don't tend to last very long. A friend of ours once noticed a sealed cardboard box jumping all on its own by the side of a road. Inside, she discovered two juvenile ferrets, which she named Jack and Vera after *Coronation Street* characters. Somehow – and inevitably – we ended up looking after them. We soon worked out that you only need to put dog leads on your ferrets and take them for a walk if you want to get the neighbours gossiping.

There's a groanworthy tip for telling weasels and stoats apart: weasels are weaselly identified whereas stoats are stoatally different. Pine martens are hard to confuse with any other British mustelids: a male weighs about four pounds and a female three, which makes them roughly four times bigger than a stoat and five times smaller than an otter. The species most likely to create uncertainty is the polecat, but that's black and white and looks like it's wearing a Zorro facemask. With an extravagant bushy tail, chestnut coloration and a creamy, sometimes even ginger, bib (or 'gorgette') under its throat, the pine marten is an altogether more graceful species. We do recognise that there can be differences of opinion over the legitimacy of such claims. Tim: 'Pine martens are the most beautiful of all the creatures that we're writing about.' Fiona: 'Don't talk wet!'

Nomenclature for martens is a shambles. The American marten, native to Canada and some (mostly western) parts of the USA, is commonly called a pine marten despite being a completely separate species. It's one of eight marten species worldwide, but only two are found in Europe. Ours is *Martes martes*, which, as Johnny Birks notes in the definitive book on pine martens, happens to mean 'Tuesday Tuesday' in Spanish;[1] the other European species is the stone marten, *Martes foina*, otherwise variously known as the beech marten, the white breasted marten and – most bafflingly of all – the house marten. House martens and house martins: it's not as if other names weren't available.

Stone martens tend to prefer warmer regions, so, unlike the pine marten, they haven't bothered moving north into Scandinavia. (That said, Poland is home to both species, and it can feel a bit parky there in winter.) Pine martens lose their mojo when it gets too hot, which is why in the Iberian Peninsula you'll only find them in a narrow band along the northern coast; stone martens extend all the way down to the Mediterranean. That still leaves a massive overlap in their ranges, covering France, Germany and Italy as well as much of Eastern Europe.

On your summer holidays, you'll do well to work out from a fleeting glimpse whether it's a marten of the stone or pine persuasion. The stone marten's face is more angular, its ears rounder and its bib whiter. Pine martens are shy and hide themselves away in woodland (not necessarily pine); stone martens are far more brazen and often turn up in roof spaces and other man-made structures. Those structures include the underside of cars. Stone martens have a proclivity for chewing through washer fluid hoses, water lines and ignition cables, having been provoked into a frenzy when another marten has got there first and left its scent behind. Cars are portable scent markers, so if you park in a second marten's territory, it'll go berserk and start destroying all traces of its rival. Johnny Birks tells us that in Germany in 2017 the cost of the stone martens' antisocial habit was 72 million euros, a twenty percent increase on 2016.[2] Mercedes has responded with its *Marderschutzanlage* – 'marten protection system' – which works by setting up an electric fence under the bonnet: 'Small martens, big damage – not with the M4700B device which offers effective protection through high voltage.' There's even a specialist marten-damage insurance policy: *Marderbisse* ('marten bites').

But these are trifling antics, and some martens dream big. A stone marten halted the Large Hadron Collider for a week in 2016 by chomping through some of its wires, and a second stone marten repeated the trick six months later as if to prove it wasn't a fluke. Sadly, neither lived to tell the tale. To add insult to fatality, one of them is now on display in the Rotterdam Natural History Museum, alongside such fellow worthies as a hedgehog that died wedged inside a McFlurry pot, a catfish surgically removed from the throat of a drunken *Homo sapiens* after being swallowed alive, and a sparrow shot with an airgun after disrupting a world record attempt by inadvertently knocking over 23,000 dominoes.

Stone martens may once have been native to Britain. A number of nineteenth-century nature writers differentiate between two marten species, and note the coloration of their bibs as the clearest evidence of their distinction. Marten experts are now scouring British museums for specimens that may confirm the presence of stone martens in this country until well into the Victorian period. It wouldn't be any great surprise given the number of what are called cryptic species hiding behind each other. After all, soprano pipistrelle bats were only distinguished from common pipistrelles in the 1990s, and it wasn't until 2010 that DNA evidence proved the existence of two distinct African elephant species. If the museum hunt were to turn up a stone marten specimen, the species would be given the same UK Red List status as the wolf: native, but extinct some time after 1500. Under those circumstances, British governments should actively consider a reintroduction programme. Don't hold your breath.

In case you're pining for more facts about martens . . .

The name for a group of martens is a richness. As martens aren't great lovers of each other's company, the word has always been more likely to refer to a collection of skins than to the living animals. You were certainly rich if you owned them. Marten pelts were being exported from Scotland in the fourteenth century, and archaeological evidence suggests that our prehistoric ancestors were already hunting and skinning them.

The Croatian currency is the kuna, which is also the word for pine marten. Their furs were once the unit of currency, so pine martens are depicted on many modern

Croatian coins. The marturina was a tax levied in the southern territories of the Kingdom of Hungary, the name (literally, 'marten's fur') deriving from its origins as a charge specifically on the trading of marten skins. To this day the Slavonian coat of arms features a pine marten *courant* – running, mid-stride with all four legs in the air.

The best way to attract pine martens to your camera trap is to invest in Mink Lure – the gooey anal gland secretions of the American mink. Mink Lure drives pine martens doolally. They'll come from miles around to roll in it. Think of the worst stench that's ever offended your nostrils, and you won't even be close.

Pine martens have a gestation period of thirty days, giving birth in the spring, usually to two or three kits. Yet they mate in the summer. This is because they practise delayed implantation, by which fertilised embryos (blastocysts) are kept in suspended animation for months. It's likely that all species of mustelid once used delayed implantation, but some have evolved out of it.[3] Our British species are divided down the middle: stoats, badgers and pine martens have delayed implantation; weasels, polecats and otters don't.

Alone among British mustelids, pine martens have semi-retractable claws. In common with the squirrels that they're chasing up and down trees, their hind feet can turn 180°. The naturalist H.G. Hurrell noticed that a pine marten's feet are 'exceptionally large for the size of the animal', with tracks the same size as a fox's. Hurrell speculated that this gives them a firmer grip, allowing them to manoeuvre more effectively when vertical.

Chaucer warned of 'The smylere with the knyf under the cloke'. The wildlife version of this sneaky killer is the pine marten, which – at least according to legend – conceals beneath its bushy tail an extra claw with which to attack other animals. And if you believe that, we have a get-rich-quick scheme involving a Nigerian prince just for you.

The most famous pine marten in modern literature is Pantalaimon, the dæmon of the heroine Lyra in Philip Pullman's trilogy, *His Dark Materials*. All children have a dæmon that can switch between animal forms until, at adulthood, it becomes fixed as an external manifestation of their personality.

George Best liked to tell a story about a hotel bellboy delivering champagne to his room. Best was in bed with Miss World and had strewn his casino winnings all over the floor. 'Ah, George,' the bellboy is supposed to have said as he surveyed the scene, 'where did it all go wrong?'

We thought of George Best while we were enjoying our own, admittedly less glamorous, bellboy moment. There we were, completely drenched on an exposed Welsh hillside in the sideways rain, kneeling down on a gravel path next to a berry-studded lump of black matter. What we were doing was sniffing poo. Where did it all go wrong? But like George Best, we were actually having a great time. Scats 'are our meat and drink, figuratively speaking', writes Johnny Birks.[4] We'd better make clear that the emphasis is on 'figuratively'. One reason for the faecal fascination is that pine marten poo comes in a variety of shapes and sizes, colours and textures. Martens have a curious habit of gyrating their back end – a little disco-dancing hip-wiggle – so that their poo gets deposited in coils and curls. Even that pattern isn't absolutely diagnostic. Pine marten poo can be confused with fox, badger, dog, polecat or otter, so the smell test is essential. Some people claim that pine marten poo smells of Parma violets. It doesn't smell of Parma violets. It does, though, have a subtle and pleasant sweetness with resiny notes; if the smell makes you gag, you've got the wrong species. The pine marten has barely any scent, which is why it used to be called the sweetmart, whereas the smellier polecat, with pungent poo to match, was the foulmart. Poor polecats are mightily abused: their scientific name, *Mustela putorius*, means 'stinky weasel'.

Poo is important for pine martens as well as the humans searching for them. Like most mustelids, martens are loners who find the proximity of another adult of the same sex intolerable. They do their best to avoid conflict by marking their territory with scats, which are usually placed in open areas: the middle of roads and tracks, elevated on tussocks, or deposited on the top of den boxes, tree stumps or logs. For good measure, they sometimes add a squirt of anal jelly. As well as being an effective KEEP OUT! sign, these secretions convey messages about the age, sex, condition and reproductive state of the depositor. This allows a lonely hearts pine marten to know when and where its advances will be welcomed. Sadly for biologists wanting to come up with population estimates, there's no easy equation between the number of scats and the number of pine martens. Much like humans, some poo a little and some a lot.

So off we went to Hafod Forest, a mile or so south-west of Devil's Bridge, in search of scats. We walked for a long time and didn't find any. There were depressingly few signs of wildlife at all: no birds, little in the way of plant biodiversity, not even as much as a badger latrine. The most exciting discoveries were puffball mushrooms and very ancient anthills. Several hours into the search, it was Patrick who spotted the first scat, by a path amidst piles of sheep droppings. When everyone had taken a turn at olfactory verification, Fiona found a stick and pushed the poo into one of her sample tubes; on we went with a new spring in our step. Never before has a poo been so enthusiastically celebrated. Yes, we know we're crazy, but it could be a whole lot worse. We could be birders.

After a lunch spent arguing over whether we'd rather see a pine marten or a unicorn – we only tackle the big questions – we made our way to another reintroduction site nearby. This was The Arch, named after a large stone structure that had once been the gateway entrance to the Hafod Estate. The mansion had fallen into disrepair and was demolished in the 1950s, leaving The Arch standing on a picturesque road to nowhere. Here was an answer to where the pine martens had gone, because we found scats every couple of hundred metres along the paths, a lot of them decorated with rowan berries. As if that weren't excitement enough, we also came across several dollops of pine marten vomit – those rowan berries again! – all of which we lovingly nudged into tubes. (Pine marten poo can look like vomit, but this was probably actual vomit: pine martens will deliberately gorge on rowan berries to clear out their stomachs.) We hadn't seen pine martens, but we had our samples and a plan.

Reintroductions are precarious for one very obvious reason: if a species has disappeared from a sizeable proportion of its native range, dumping a few animals back into the same depleted landscape and expecting a different outcome is unlikely to work. Hope isn't a strategy. Most reports detail only successful operations, on the basis that no one wants to draw attention to a failed project, but estimates of success based on the *available* data range from twenty-six to sixty-seven percent.[5] So the best-case scenario is that about a third of projects fail, and buried in these numbers are serious welfare issues. You really don't want to be an animal in the front line of a reintroduction project. Captive-bred individuals can be ill-equipped to cope with their freedom, while translocated animals may have particular local adaptations that are unsuited to their new environments. If they're a prey species, they'll have no experience of evading predators; if they're predators, they may prove hopeless at catching prey. All will be extremely stressed by their sudden change in circumstance, and many will die in short order.

Britain doesn't have much experience of reintroductions, thanks to a succession of governments that have been either openly hostile or forbiddingly bureaucratic (and usually both). This serves as inspiration and justification for the so-called 'maverick rewilders' who take matters into their own hands: as the unexpected beaver renaissance shows, facts on the ground – or in the river – trump any number of consultation documents, feasibility studies and interminable pilot schemes. While official organisations have been busy filibustering projects out of existence, a network of shadowy individuals is illegally releasing animals across the British landscape. To be honest, they're not all *that* shadowy, but – like Macavity the Mystery Cat – they always have an alibi and one or two to spare. For example, nobody seems to know anything about an informal transfer market several decades ago that definitely didn't exchange polecats from the Welsh borders with pine martens from Scotland.

The choice is usually between illegal releases and no releases, but not every conservationist approves of these undercover activities. As one Twitter response to a sympathetic *Guardian* article put it, 'Maverick Rewilders? Or disorganised, poorly planned and random releases of captive-bred species?' In fact, there's no point releasing pine martens on the sly, because you need a few dozen female martens to give yourself any hope of establishing a sustainable population. Even leaving aside the practical prospects, sourcing animals from far-flung locations can mean that the newcomers are genetically ill-suited to their new home. There's also a massive risk of inadvertently introducing disease. As we've seen with wild boar, those risks aren't merely theoretical: the spread of African swine fever owing to the introduction of pigs into sub-Saharan countries is an object lesson in what can go wrong. Mavericks working out of a garden shed simply don't have the know-how or resources to screen for disease.

After centuries of persecution, pine martens got lucky. Their big break came in the 1990s when their cause was taken up by the Vincent Wildlife Trust, a mammal conservation charity with the budget, patience and scientific expertise to overcome even the most inert of government bureaucrats. Too many conservation projects opt for a quick fix and, after milking the media attention, hurriedly move on to something else. The VWT keeps monitoring long after everyone has gone home. It initially opposed pine marten translocation projects altogether, because it worried that the genetic distinction of English and Welsh martens would be eradicated by the arrival of their Caledonian cousins. So it went in search of pine martens in Wales, and found almost nothing. Only after several years of depressingly blank fieldwork was the VWT satisfied that there were no English or Welsh genes left to save: martens, it concluded, had been functionally extinct in England since about 1925, and in Wales since 1950. The occasional marten still showed up from (ahem!) who knows where, but there was no evidence of any sustainable breeding population. Bring on the Scots!

The VWT could afford to be in no hurry because pine martens were doing fairly well in Scotland and Ireland, and were beyond all hope in England and Wales. What followed was a decade of shuffling inexorably forward, answering the conservation questions one by one. How many pine martens should be translocated, and over how many years? Where should they be caught and released? How would they be monitored across their new ranges? Just as important was the issue of hearts and minds – or as the paperwork likes to call it, stakeholder buy-in. If martens had already been wiped out by habitat loss and human persecution, why should they fare any better second time round?

In some quarters, rewilding has become toxic. A newspaper article summed it up succinctly in 2020 when taking a passing swipe at 'a trendy form of eco-friendly land management known as "re-wilding"', with the scare quotes and hyphen making the entire movement look even more outlandish. Often, rewilding is seen as a posh middle-class pursuit, imposed on rural communities by sanctimonious townies with no interest in the people who, rightly or wrongly, worry about threats to their livelihoods. In this respect, the VWT played an absolute blinder. It settled an employee into accommodation near Devil's Bridge, where for two years he built relationships and spent time talking and listening as the public face of the project. Trust was earned meeting by meeting, school visit by school visit. Locals could rest assured that there'd be no nasty surprises, because this was a genuine consultation: the project wouldn't go ahead without overwhelming support.

If the language of rewilding is taboo, so are several related words. It's no accident that the restoration of pine martens was achieved through the Pine Marten Recovery Project, not the Pine Marten Reintroduction Project. This preference for words like 'supplementing' and 'reinforcing' was shared by Welsh government agencies, which must also have liked the appeal to patriotism and localism. Forester Huw Denman spoke regularly at public events and in the media, emphasising the evidence for a long tradition of martens in Wales found in songs and poems, place names and historical records. For example, the Welsh for pine marten is *bele*, hence Bele Brook in Powys and several others. The message from all of this enthusiastic public engagement was that pine martens have belonged in Wales since time immemorial.

The most important fear to address was about pine martens killing sheep. This isn't a debate with grey areas. Pine martens don't kill sheep – not even a lean, mean and hungry pine marten chancing on a frail newborn lamb. Countryside myths can be more pernicious than their urban counterparts, and many sheep farmers have heard stories about martens ripping the throats out of ewes and dragging carcasses into burrows. One fevered version of particularly ancient lineage insists that pine martens will grab a sheep by the nose and start eating inwards.[6] It's hard to counter these unhinged imaginings with boring facts, especially when some grizzled old-timer stands up at a meeting and insists that he once saw it happen in his youth. The VWT calmed nerves by introducing a compensation fund to which farmers can apply whenever a sheep is shown to have been killed by a marten. Guess how many cases there have been to date.

In the end, the lovebombing worked: ninety-one percent of locals supported the reintro . . . sorry, the *recovery* of pine martens in Wales. It's still fascinating to hear the bitter-enders' concerns in the various vox pops recorded by the VWT.[7] There's the curmudgeonly desire neither to have one's cake nor eat it: pine martens won't attract tourists, and if they do, well, we don't want tourists anyway. Several respondents expressed the more sympathetic fear that pine martens would have a damaging effect on other wildlife: 'The wood pigeon is coming back strong from nothing. Something like this is going to have the nests of them. What you're gaining in one way you're losing in the other, and personally, I'd rather have the pigeons.' Wait, what? Wood pigeons? The wood pigeon is, by some margin, the most common large wild bird in Britain, with an estimated 5.1 million pairs, and populations have been increasing steadily since the 1960s. A handful of pine martens won't put much of a dent in those figures. And what if pine marten numbers get out of control? As Tim's nonagenarian grandma always says, 'Chance'd be a fine thing!' Pine martens have low reproductive rates and large territories. We aren't ever going to be up to the eyeballs in them.

You might reasonably expect that the Pine Marten Recovery Project would win the steadfast commitment of government agencies. After all, pine martens were listed among 'species of principal importance for the purpose of maintaining and enhancing biodiversity' in the 2006 Natural Environment and Rural Communities Act. So much more perplexing, then, that government organisations not only failed to encourage the VWT, they almost succeeded in scuppering the entire scheme at the last minute.

The recovery programme needed to navigate between the Home Office and the Welsh government's environment department. Scylla and Charybdis . . . a rock and a hard place . . . the devil and the deep blue sea: there's no phrase capable of capturing the full horror. Home Office involvement was required by law because the VWT was collaborating with a research institute – in this case, the University of Exeter – to find out critical information about pine marten behaviour. Only with the ministry's blessing could the animals be anaesthetised and fitted with radio collars. This is because any work with animals conducted for a scientific purpose is regulated by the same laws and processes designed to regulate vivisection. Cut off a young ram's testicles without an anaesthetic in order to monitor new ways to alleviate pain and that's regulated because you're carrying out scientific research. Go ahead and do it because you're a farmer who doesn't want the vet's bill and . . . nothing. Marking an animal by injecting a microchip isn't regulated, whereas making the same size of hole to acquire a sample for genetic analysis is. Deliberately or not, the fees and administrative burdens imposed on researchers undertaking relatively mild interventions act as a tax on conservation, punishing the organisations that behave responsibly and encouraging others to work secretly and carelessly.

Time ticked by, applications for the work were submitted, resubmitted and re-resubmitted, but answer came there none. (Fiona can sympathise with the VWT: it recently took her well over a year to transfer her Home Office licence to a new university and get permission to carry on doing the same work she's been conducting for more than two decades. Best not to get her started on the yearly torture of renewing her Natural England licence.) Not wanting to miss out on the fun, the Welsh government's chief veterinary officer then started to worry that pine martens might be vectors for disease, passing on bovine tuberculosis to cattle. For a while, it looked like the project would be prevented from releasing martens at any site owned by the Welsh government. Never mind that there hasn't been a single case of pine martens carrying bovine TB anywhere in the world, nor that they spend their time in trees rather than in fields with cattle. (To prove the point, in Irish they're known as 'an cat crainn' – the tree cat.) As so often with government vets, they were demonstrating the truth of Maslow's Law: when you only have a hammer, you treat everything as if it were a nail. There followed some desperate last-minute pleading, and urgent phone calls between the VWT and the various officials. Permission was finally granted on the morning that trapping was scheduled to begin.

One crucial requirement when preparing a landscape for the arrival of pine martens is to provide somewhere that they can call home. Pine martens are arboreal, but despite their name they seem to show no preference for conifer over deciduous woodland. They can even survive in rocky areas with no woodland at all. These formed some of their last refuges in Scotland; it's harder for humans to hunt them up a mountain. Still, left to their own devices they much prefer tree cover. Old-growth woodland is their favourite habitat because it provides a smorgasbord of beetles and berries and birds but also, critically, offers a multitude of natural denning sites in the trees. That puts them at a serious disadvantage on this side of the Channel, because commercial conifer plantations are hopeless, and most other British woodlands are so thoroughly over-managed that there are no old or decaying trees left for suitable dens.

In an effort to rectify this problem, the VWT started testing artificial den boxes in Scotland's Galloway Forest Park in the early 2000s. After showing initial promise, the project was continued by pine marten experts Johnny Birks and John Martin. Together with a team of volunteers, they valiantly lugged fifty thirteen-kilogram wooden boxes around the forest, fixing them at a height of about four metres above the ground by means of an ingenious system of ladders, ropes, pulleys, and probably circus skills. Careful monitoring demonstrated that the boxes were a resounding success. Whereas, previously, almost all martens denned on the ground (leaving the kits vulnerable to predation by foxes and golden eagles, and disturbance from people), now they positively leapt into their new tree-house accommodation. Between thirty and seventy percent of the den boxes became occupied every year during the twelve years of follow-up monitoring, and they were regularly used for breeding.

One issue remained. In precisely those areas where the boxes were most needed – immature and heavily managed woodland – there were few trees sufficiently sturdy to support them. Hence the design of the 'Galloway Lite' den box, which comes in at about a third of the weight. It's based on a blue plastic barrel adapted for pheasant feeders, and, as a pine marten estate agent might say, it provides durable, spacious and waterproof accommodation. In one Scottish project, between a third and three-quarters of the forty-seven boxes installed were occupied in any given year, and, after three years of monitoring, ninety-four percent of the boxes had been used. With all this evidence in their favour, similar boxes have now been strategically positioned across the Welsh countryside, awaiting the arrival of their new tenants.

The reintroduction team had a policy of never taking more than four pine martens from each site. There's no point introducing them to Wales if you unsustainably drain the original Scottish populations in the process. We know that numbers are holding up: DNA sampling in Scotland suggests no reduction in genetic diversity or population size, and pine martens are continuing to extend their ranges. Of the pine martens captured by the VWT, any that looked a bit decrepit were released as ineligible, as were juveniles born that year. What the project needed more than anything were adult females that had already mated in the summer. Those healthy martens were then put into a specially adapted dog van, given plenty of blueberries to keep them happy, and driven down to their new Welsh homes.

Once fitted with radio collars, the martens spent a couple of days in special pens designed by Chester Zoo. This gave them the chance to acclimatise themselves, although one particularly impatient marten decided to dig its way out. A network of VWT staff and volunteers monitored the animals after release, using a combination of radiotracking, camera-trapping and good old-fashioned ear-to-the-ground intelligence. There was the usual mixed picture that you'd expect from even the most professional of translocation schemes. One marten died while being prepared for release; at least one was killed by foxes, which don't eat them but aren't sentimental about competitor species; one was shot by a landowner who claimed that it had been raiding a red kite's nest; two others died of infections; several more disappeared without trace, possibly because none of the radio trackers ever got close enough to pick up a signal. One, as we know, sailed into the USA with her radio collar intact. Another hung around for six months, but then disappeared and was spotted on a security camera more than 100 kilometres away in North Wales. It disappeared again and was eventually found dead on a road in Staffordshire. Whether it had migrated there on its own, or been given a helping human hand, we don't know. Fortunately, these troublemakers are vastly outnumbered by the quiet ones who don't cause a scene. As a result of the releases, pine martens have been born in Wales every year since 2015. It'll take many more years of tracking their populations before we can be certain that they're established, but the early signs are positive.

Regardless of the reintroduction programme in Wales and England, pine martens are on course to cross the border from Galloway into Kielder Forest in Northumberland under their own steam and start breeding over the coming decade. We know from camera-trapping that pioneers have already arrived. From there it's just a moonlight flit to the North Pennines, then the North York Moors, the Lake District, the Yorkshire Dales and the Forest of Bowland. The entire process will speed up immeasurably if – as seems to be on the cards – funding comes through to support the release of Scottish martens in south Cumbria. Pine martens are also being reintroduced by Gloucestershire Wildlife Trust into the Forest of Dean, where they can keep the wild boar company; they should join up with the Welsh martens in about ten years. Dartmoor and Exmoor are potentially next, with a stakeholder consultation starting shortly. The other area that looks promising is near the Devon–Somerset border. There we have a personal interest, because the VWT's computer modelling shows that after a few years our back garden will become a pine marten hotspot. Well, to be finicky about it, each of the map's pixels covers a much larger area than that, but we'll buy Mink Lure by the barrel if we have to.

On the principle that their enemy's enemy is their friend, foresters are big fans of pine marten recovery programmes. Martens eat grey squirrels whenever they can get hold of them, and if there's one thing foresters can't stand it's grey squirrels. Their habit of chewing bark destroys the commercial value of trees; what might otherwise have been suitable for furniture is downgraded to firewood and chipboard. News from Ireland and Scotland that grey squirrels seem to be disappearing from areas recolonised by pine martens has therefore persuaded foresters that they'd like nothing more than to share their woodland with their newly discovered ally. Better still, the latest evidence suggests that red squirrels are recovering in areas of Ireland where greys have declined, providing the best of all possible worlds: more pine martens, more red squirrels, fewer grey squirrels, and deliriously happy foresters.[8] It sounds too good to be true, so what's the catch?

Pine martens are seriously hardcore. They'll eat pretty much anything, from fruit and insects to birds, frogs, toads, small mammals and carrion. They'll even chew through the nests of wasps and bees when they're feeling peckish. Field voles – which can reach very high densities in sections of conifer forest with young or recently felled trees – form up to eighty percent of the pine marten's diet in Scotland. Voles are naturally absent from Ireland, so other items are on the menu there. When it comes to squirrels, pine martens aren't fussy: both red and grey flavours are acceptable. These seem to be seasonal delicacies, eaten mainly in spring and summer and almost never in autumn or winter. Nobody has yet assessed the age of grey squirrel remains found in pine marten scats, but the seasonal bias is evidence that they're most likely to be naive juveniles, or kits raided from nests.

To say the least, it's an unusual strategy: saving a species by introducing one of its predators into the landscape. If pine martens enjoy tucking into juicy young squirrels, it doesn't seem likely that reds will recover thanks to their presence. Yet there are several interlocking reasons why this may nevertheless be happening. All those thousands of years of co-evolution with pine martens have made reds much savvier than greys: they're lighter and more nimble, they tend to nest in smaller nooks, and their juveniles keep an eye out for predators among the branches. If for some bizarre reason you want to frighten a red squirrel, take some pine marten scat, mix it with water, and spray it around a nut feeder. The reds flick their tails, look around nervously, and find somewhere else they'd rather be. Try the same thing with greys and they carry on without a care in the world.[9] In their native range, grey squirrels have no tree-climbing predators, so for them the Spanish Inquisition strikes, dressed as a pine marten, when they least expect it.

It isn't far-fetched to imagine a native predator feasting on a non-native prey species. (As the next chapter's tragic tale of water voles and American mink illustrates, the reverse is common enough: non-native predators wiping out a native prey species before the latter has a chance to understand what's going on.) But it's also possible that the process will rapidly result in grey squirrels getting shrewder. Moose in Yellowstone National Park developed hyper-vigilance in response to reintroduced wolves in a single generation, and northern brown bandicoots and common ring-tailed possums, previously hammered by the introduction of non-native red foxes to Australia, have learned to recognise foxes for the predators they are. We own a book called *Outwitting Squirrels*, which goes to great lengths to explain how to protect bird-feeders from the ingenuity of greys. There's even a recent scientific paper demonstrating that grey squirrels are extraordinarily fast learners.[10] Some of them may get eaten in the meantime, but we wouldn't bet on grey squirrels being outsmarted by pine martens for long.

Across Europe, red squirrels don't turn up especially often on a pine marten's *carte du jour*: a recent review found that their remains were present in four percent of pine marten scats (with a range of 0–19%).[11] By comparison, the only available studies of grey squirrels (one from Northern Ireland and one from the Republic) found them in twelve percent and sixteen percent of scats respectively. But before we get carried away and say that pine martens can never be important predators of red squirrels, we should remember that they're opportunistic hunters. Red squirrels feature if they're locally abundant and other prey are scarce, such as when natural cycles send field vole populations crashing. The latest research from Northern Ireland also highlights the vulnerability of red squirrels to pine martens in non-native conifer plantations that lack the structural complexity of native woodland. So while grey squirrels are always suppressed by pine martens, impacts on red squirrels seem to be habitat-dependent.

One fashionable theory doing the rounds, from newspaper articles to polemics on the restoration of Yellowstone National Park, suggests that the reintroduction of native predators creates a 'landscape of fear'. The general principle is that while predators may not depress prey populations through hunting alone, their very presence triggers behavioural changes that affect prey population dynamics. If you restore an apex predator, prey species that would otherwise be stuffing their faces, having lots of sex and producing endless offspring are now obliged to spend energy avoiding becoming lunch. Conservationists desperately want the theory to be true: let's reintroduce wolves to Scotland and solve the deer problem! So it's painful to admit that the evidence for this Panglossian concept isn't always as straightforward as we might hope. Deer in Scandinavia, for example, are certainly eaten by the lynx and wolves recolonising the region, but there's little evidence that their distributions or behaviours have changed.

Studies of grey squirrels in the Welsh reintroduction area illustrate the difficulties closer to home. Some clever research was carried out using GPS radiotracking to monitor concurrently the behaviour of twenty-nine adult grey squirrels and the first twenty reintroduced pine martens.[12] The squirrels responded almost immediately to the predators' presence, and their home ranges and distances travelled were substantially larger where there was greatest exposure to pine martens. This was the opposite of the team's prediction that home ranges would shrink owing to increased vigilance and reduced foraging. The researchers suggest instead that the increased movement and home range size could be caused by altered competition with other unstudied squirrels (involving, potentially, the opening up of new territories after pine marten predation). Alternatively, they argue, fear could still be playing a part: some species will limit the amount of time they spend in any given place when they fear an attack, with the consequence that they travel further and have larger home ranges. But the squirrels didn't just up sticks and move: their home ranges remained centred in the same places. This is hardly the behaviour anyone would expect from a frightened animal. And if grey squirrels are affected by this landscape of fear, how do they receive the memo about predators moving in? The nut-feeder experiment shows that spraying pine marten scats around the place doesn't seem to induce terror. There may be a more direct effect if squirrels spot a lurking marten, but how often does that happen? And why would a marten be scarier than, say, a man with a gun or a noisy terrier – which many grey squirrel populations encounter without any suppressive effect on their reproduction? It's proximity to humans rather than natural predators that explains most of the vigilance behaviour of elk studied in

Canada,[13] while in Poland's Białowieża Primeval Forest the distribution and behaviour of wild boar and deer are driven mainly by hunters and not wolves. Could it be that grey squirrels are already in a landscape of fear, surrounded by people, but they're thriving regardless? Their behaviour apparently *does* change in places where there are pine martens, but we still need to work out why and with what outcomes.

As for whether pine martens can eliminate grey squirrels and save the reds, the answer is tentative but important: perhaps in some places. The most recent models suggest that in Scotland, northern England, Northern Ireland and the Republic of Ireland, where red squirrel populations are still significant, the effects are beneficial. The government agency Forestry and Land Scotland is currently installing den boxes in the path of advancing grey populations, hopeful that the pine martens will move in and stem the tide. The prospects are less promising in Wales and southern England, home to vast extant populations of grey squirrels and, in many areas, very limited habitats suitable for pine martens: there's simply too much intensively managed farmland and urban sprawl. Grey squirrels will keep emigrating from the towns and colonising countryside spaces. A study of sixty-one individually marked grey squirrels at the pine marten release site in Wales found no increased mortality, even in the places where martens were most active.[14] It may be that the pine martens would make a bit more effort if other food sources weren't available, but squirrels are hard to catch and most of the time there are easier pickings.

*

In contrast to the general enthusiasm for pine marten restoration in England and Wales, the Scottish Gamekeepers Association and estate managers, supported by the Game and Wildlife Conservation Trust, periodically call for exemptions to the legal protection for pine martens. They'd like to start killing martens again, or at least to move them a long way away. Surely no one should doubt the sincerity of their stated motive: these organisations are inspired purely by their altruistic love of the capercaillie – a ground-nesting wild grouse as absurd as it's magnificent. The word 'capercaillie' means 'horse of the woods' in Gaelic, which only tells us that those Gaels must have been smoking some good gear. Imagine, instead, a pint-sized Emu with no Rod Hull attached. The male capercaillie becomes insanely territorial in the breeding season, attacking anything that moves, including hapless walkers and cyclists. Unfortunately, their feisty attitude can get them into heaps of trouble. There's footage online of two capercaillies fighting, utterly oblivious to anything around them; a golden eagle swoops down and kills one on the spot, while the other, instead of retreating and thanking his lucky stars, decides that the really clever thing would be to attack the eagle – with sudden, bloody and entirely predictable consequences for the capercaillie. Geniuses they ain't.

Adapted to the cold, the capercaillie is common in conifer forests across Scandinavia, Russia and Romania, whereas in Scotland it's classified as Endangered on the regional IUCN-compliant Red List. So it's struggling here, and not for the first time. Despite the Gaelic name, archaeological evidence for the capercaillie's presence in Scotland is much weaker than for

England and Ireland, and no pre-1500 written record exists. Following its extinction in Britain and Ireland around 1770 – probably owing to deforestation and over-hunting – the capercaillie was reintroduced to Scotland from Swedish stock in 1837. It flourished until the mid-1970s when there were about 10,000 pairs, but then began a sustained decline that now threatens its survival for a second time. About three-quarters of the remaining capercaillie population is restricted to forests in Badenoch and Strathspey in north-east Scotland. A survey in 2009 suggested that the population had fallen to 1,268 birds – down by a third from the previous estimate in 2004. Counts at lekking sites (where males gather to display and fight) indicate that numbers have fallen still further since that time, with around 700 birds remaining.

The survival and productivity of capercaillie populations are influenced by many factors. Of these, the most important appear to be habitat and weather. The capercaillie prefers conifer forests with open canopies, a rich internal structure, and dense ground cover with abundant shrubs from the blueberry family; in Britain, these are usually bilberries, which also happen to be much loved by pine martens. (Lancashire born and bred, Fiona calls them whinberries, and they have various other dialect names in different regions of Britain.) Historically, the Caledonian old-growth forest of Scots pine would have been ideal, but now less than one percent of that forest remains. To add to the problems, wetter and colder Junes, which are part of a long-term weather pattern in Scotland, result in soggy chicks, low availability of caterpillars for them to feed on, and poor survival rates. In some areas, winter disturbance by hill walkers and skiers affects survival, as do birders traipsing through woodland in search of their lekking sites in the spring. Capercaillies also get themselves killed at remarkable rates by flying into deer fences and wind turbine masts. Then there's the bit that excites gamekeepers: predation of eggs and chicks by foxes, corvids and other predators.

The proponents of pine marten culls insist that martens are thriving in Scotland, while capercaillies are so vulnerable that any impact from predators can't be allowed to continue. There are already extensive culls of foxes and corvids to protect capercaillies: why not add pine martens to the kill list? As one particularly vociferous culling proponent puts it in shouty capitals, 'CAPERS ARE IN A PREDATOR PIT!!!!!!'. But no one is denying that pine martens eat the capercaillies' eggs and chicks. In Abernethy Forest, where much of the remnant capercaillie population is found, a third of nesting attempts will end up providing snacks for pine martens. This seems a lot, but it's also about the same as average nest loss rates elsewhere in Europe.

Pine martens and capercaillies have coexisted across their range for 12,000 years, during which time capercaillie populations have bred successfully in the presence of pine martens and countless other predators. That in itself should be a clue that the pine marten's role is doubtful, and a series of research projects have come to the same conclusion. A 1995 study in fourteen forests found no link between capercaillie breeding success and a pine marten abundance/activity index.[15] A further study of the impacts of various predators on capercaillies in north-east Scotland confirmed that, while signs of pine martens were more abundant than in 1995, there was no evidence to suggest that martens are affecting capercaillie breeding success. The remaining strongholds for capercaillies in Scotland are also those that have key pine marten populations. That's not surprising. If the capercaillie is indeed native to Scotland, it existed alongside northwards of 50,000 pine martens during the Mesolithic period. It's ridiculous to blame the martens when there are only 3,000 of them left. What affects capercaillies most, as various studies have shown, is the weather; no quantity of dead martens will change that.

Unperturbed, every now and then the Scottish Gamekeepers Association and its allies advocate a 'pilot' cull. Killing a few martens to find out whether we should kill a whole lot more sounds like useful research if you're that way inclined, but the maths demonstrates that, even on its own terms, it's a senseless strategy. The beleaguered mother capercaillie currently manages to raise two chicks out of every five nesting attempts. Let's assume, optimistically, that culling martens increases that rate to three chicks. It would require a study of 198 trial nest sites where predation is prevented, and another 198 control nest sites, just to be confident that the change really was owing to marten removal and not just chance fluctuation. Culls on such a massive scale would devastate pine marten numbers, which means that no pilot cull would ever generate the required information without massacring an endangered native species in the process. We can only conclude that we're missing the point. After all, there are plenty of so-called 'scientific' culls – of wolves in Scandinavia, for example, or whales in the north-west Pacific and Antarctica – where the science provides a fig leaf for ulterior motives.

Recently, the Scottish government commissioned an expert review of what should be done to rescue capercaillies. To no one's surprise, that team of experts didn't include a single mammal conservationist. It was a classic case of picking the judges to ensure the verdict: lo and behold, the report concluded that pine marten recovery may be playing a role in the capercaillie's demise. The Chair of the Scottish Gamekeepers Association claimed to be vindicated, and looked forward to 'sav[ing] this iconic species' after years of 'misspent public money' wasted on pine martens. What he didn't highlight was that even this report worried about the extent of predator control that would be necessary to achieve 'a measurable impact on the Scottish capercaillie population'.

Not wanting to advocate killing an endangered native mammal, NatureScot was immediately at pains to point out that 'In the case of pine marten, this [control] would be non-lethal, through trap and release as part of reintroduction to other parts of the UK.' Transportation to another country won't work if it's only applied to ten, twenty or even a hundred miscreant pine martens. So are we really advocating the creation of a landscape devoid of all native predators, including not just pine martens but also the thousands of badgers, foxes and crows that'll have been more summarily dispatched? The intervention would be so huge that, in effect, we'd be creating a safari park. If that's really what people want, why not just put the capercaillies in a zoo and be done with it? It'd be cheaper, and a lot more ethical.

Concerns about negative impacts of pine marten restoration aren't limited to capercaillies and game birds. In the Forest of

Dean, the Forestry Commission's plan to reintroduce pine martens was almost derailed by a failure to take into full account the potential impacts on bats. The VWT arrived to save the day by fitting pine marten excluders to horseshoe bat summer roosts. We have to keep our fingers crossed that the martens don't start taking hibernating bats from the disused mines scattered around the region; the entrances to those sites are simply too large and complex to be defended against a determined marten. The abandoned army tunnels of Nietoperek in Poland provide homes for some of Europe's largest aggregations of hibernating bats; stone martens and pine martens in the region do snack on the occasional bat, as we can tell from DNA in their scats. Even so, the scale of the hunting has never reached levels high enough to trigger alarm. Most British bats squeeze into tiny crevices, and artificial bat boxes have narrow entrance holes designed to keep out birds and other unwelcome guests. It's unlikely, then, that pine martens will pose a threat to the woodland bat community.

As for birds, pine martens raid their nests for chicks and eggs. It's always wise to plan for a rainy day, and females in particular will take advantage of the springtime food glut by caching hoards as an emergency supply to feed their hungry kits. Birds that tuck away into holes (for example, the tit family) are less vulnerable than ones that nest in the open (such as the thrush family). Nevertheless, even the tits aren't completely safe: not dissuaded by the fact that their body won't fit through the gap, martens grab eggs and chicks with their paws. In response, blue tits build shallower nests in locations with high predation rates, so as to increase the gap between the entrance hole and the nest cup. These nests are suboptimal in other respects; they're less effective at protecting the nestlings from waterlogging. Work in Poland shows that parent blue tits revert to tall nests as soon as they feel safe from martens. Researchers can lend a hand by adding a simple piece of plastic plumbing pipe to the entrance hole of artificial nest boxes, making the distance too great even for Mr Tickle.

About that plan we mentioned a while ago . . . Roughly twenty miles south of Devil's Bridge – double that distance if you're driving – there's a hide in the middle of some woodland. To get there, you head for the back of the back of beyond and keep going. Don't bother trying unless you've got a four-by-four, and don't bother trying even in a four-by-four between December and March. 'Remote' doesn't do the place justice. On the other side of the valley, the sixteenth-century 'Welsh Robin Hood', Twm Sion Cati, once made his home in a hillside cave. Locals tell how he robbed from the rich but somehow forgot the bit about giving to the poor. Whatever his crimes, he knew that Elizabethan law enforcement wouldn't be crazy enough to go looking for him up there.

We parked our hire car on the nearest piece of tarmac by the banks of the River Towy, and there we met local volunteer Hugh Gillings. Hugh maintains the hide and records pine marten activity. Most of the pine martens released at Devil's Bridge had spread north, but a few went south, and one – let's call her Godot – ended up here on the edge of a Sitka spruce plantation. Hugh had seen her only three times in an eighteen-month period, despite putting in regular overnight stints with sleeping bag and Thermos flask. His camera trap by the hide had proven more successful. Trained on a small oak ten metres away, it had picked up foxes, deer, wood mice, countless birds and, on more than a dozen occasions, a beautiful pine marten. The marten liked that spot, and with good reason. Every so often, without warning, the oak would secrete jam and peanut butter through its bark, and would start fruiting eggs in its crannies and crevices. The marten would diligently stash the eggs one by one – we still don't know where – and lick the tree clean. Not surprisingly, even if she disappeared for weeks on end, she always came back to her magic tree.

We followed Hugh over the streams, along the narrow tracks and up the hills to the hide. At the site, we checked the latest footage – and this is where it started getting exciting: the pine marten had put in an appearance that very morning in full daylight at eight-thirty, as well as on the two previous mornings around the same time. We helped Hugh to balance the eggs and smear the peanut butter, and tried to stay nonchalant. Setting off for home, Hugh told us that if the marten turned up we must let him know, 'even if only for the devilment of it'.

Should we stay where we were, taking advantage of the remaining light and hoping the pine marten would immediately find the bait irresistible, or should we head back and get into position before dawn the next morning? Partly because we were miles from anywhere, with no phone signal and no hotel booked, and partly because we thought that the pine marten may have been bothered by all the daytime disturbance, we decided that our best chance was to return in the wee small hours. So off we went to Llandovery, the nearest town of any size but still thirty-five minutes away by car. We set our alarm clocks and dreamed ourselves amidst a richness of martens, dancing through the leaves.

The first challenge next morning was to find the hide in the dark. Satnav couldn't help us: it didn't even believe that there were roads. Then we had to accomplish the considerable feat of walking the last few hundred metres in pitch black and complete silence without breaking an ankle. It was still nautical twilight when we reached our destination, and after what can most diplomatically be described as a whispered kerfuffle ('Dad, you're the first person in history to bring an attaché case to a wildlife hide!'), we settled down and waited for Godot.

You know that feeling when you're confident of a favourable outcome – an exam or a job interview for which you've meticulously prepared? Then something unexpected happens, and the first doubt starts to chew into your stomach. Before long, you're living a full-blown anxiety dream. Well, there we were, staring through the crumbling darkness at the oak tree, and at first we were happy even though we could barely make out its silhouette. As the light improved, we became marginally unhappier, straining to spot the white shells of the eggs against the dark trunk. Our next trick was to persuade ourselves that it was still too gloomy to see them properly. We must be misremembering: we'd obviously put them on the far side of the branches, out of view. But dawn was cruel and relentless, and, as the minutes trickled past, what it began to reveal was a starkly eggless tree with no trace of jam or peanut butter on its limbs. There was just enough hope to stop us crying with frustration: the pine marten had come back on previous mornings, sometimes several days after clearing out the treats. So we sat and we sat and we sat for more hours than we care to recall, by now lost in the bargaining stage of our grief. We did get to watch a gorgeous nuthatch, and a great spotted woodpecker that kept returning to a tree about five metres from the hide. But with due respect to nuthatch and woodpecker, they're not pine martens. And Godot didn't show up.

FOUR

Water Voles and Earth Hounds

So here we are, standing in the midst of the perfect water vole habitat. Look around and you'll see a gentle stream with plenty of bankside vegetation for food and cover. The birds are singing; the dragonflies are dancing; the breeze is rustling the reedbeds. It's like something out of Enid Blyton, this landscape of lost childhoods and glorious summers past. All we need are Aunt Fanny's best scones and lashings of ginger beer.

Well, good luck to anyone trying to find a water vole that old-fashioned way. Now you can walk hundreds of miles along Britain's waterways without so much as a glimpse. When we lived in Oxford in the early 1990s, we'd wander silently up and down the canal, counting the distinctive plops as voles took to the water in case we were hungry predators. Sometimes on an outing we'd see five or six and hear a dozen more; whatever the time of year, and no matter how busy the towpath, we were sure to spot at least one. Twenty years and several house moves later, we went back to Oxford and found nothing. At some point, the banks had been dug out, tidied and reinforced with metal piling. No doubt that's handy if you own a canal boat, but, if you're a water vole, not so much. They can't dig burrows through metal.

Once upon a time, people were knee-deep in water voles, which were by far our most populous mammal. Based on archaeological evidence, Don Jefferies has come up with an estimate of close to 6.8 billion water voles in Britain during the Iron Age. At first glance that seems wildly implausible, so let's assume for a moment that the number has been overstated by a factor of a thousand, and that we should be talking about millions rather than billions. Even in this scenario, the relatively recent collapse in population size will have been nothing short of apocalyptic. Over the past forty years, water voles have disappeared from more than ninety percent of their native range, and the mere million that survived in 1995 had shrunk to 132,000 by the time of the 2018 review. That's why water voles have managed to earn – against stiff competition – the title of Britain's fastest-declining mammal; they're classified as Near Threatened in Scotland, Endangered in England, and Critically Endangered in Wales. The trend continues downwards, despite massive conservation efforts. You may see one in your local river, just like you may win the lottery. You increase your chances by visiting a site where a wildlife trust has reintroduced water voles and made efforts to get rid of the invasive American mink, but you'd better move fast: new colonies have an alarming tendency to vanish without trace if you turn your back for even a moment.

So much for the counsel of despair. The better news is that water voles can still be found, but you're almost certainly looking in the wrong place. In fact, if you want to guarantee success there's really only one right place. A tiny area of Britain, just five or six square miles, is probably home to more water voles than the whole of Wales. Covid restrictions had scuppered our plans to visit, but then the perfect excuse arrived in the form of a collaborative project between Fiona and the University of Nottingham. The job was to collect samples from wildlife and screen them for viruses, so as to make sure that those wretched humans hadn't been infecting other species as well as each other. We urgently needed water vole poo and mouth swabs, so we armed ourselves with hay and apples, as well as (obviously!) a handsaw, drainpipes and some Pringles tubes, and set off for that watery rural idyll.

Except that the East End of Glasgow is neither watery nor rural, and although we're rather fond of the area, we'd be stretching credibility if we called it idyllic. Water voles have chosen to take up residence in some of the most socially deprived districts in Britain. They live next to playgrounds, in cemeteries, on the edges of small municipal parks, in school fields, on roadside verges, near bus shelters, and in overgrown grassland between high-rise blocks. That such a rare and declining mammal should turn up here, of all places, makes no sense. More than that: it feels like a miracle.

But hang on – how can you have water voles without water? 'The water vole in Great Britain,' Fiona writes in the 2018 population review, 'is primarily riparian, usually occurring within two metres of water.' So they don't just saunter down for a dip once a day and then clear off across a couple of fields back to their burrows. They live beside and in water, and they make the entrances to their burrows under the waterline because it's their best protection against native predators like stoats and weasels. As a strategy, that worked fairly well until the semi-aquatic American mink showed up. Whereas an otter is too large to follow them into their burrow, a female mink can just about squeeze its way through the entrance. Mink do more than snack opportunistically on the occasional vole caught during a hunting expedition; they keep returning to a colony and entering the burrows to take adults and pups until the entire population is eradicated. Mink aren't the only reason for the water vole's disappearance, but they deliver the *coup de grâce* for populations already severely diminished by habitat loss.

Glasgow does indeed boast a few relatively modest colonies of riparian water voles. If you've never visited, you may be surprised by the extent of the greenery: the massive Seven Lochs Wetland Park, for example, proudly declares itself to be Scotland's 'largest urban heritage and nature park'. Although still a work in progress, it's already impressive enough to provide a model for what Britain's urban developments can achieve with a modicum of what George Bush Senior once admiringly called (while acknowledging that he didn't possess it) 'the vision thing'. What's strange about the water voles is that most of them ignore these natural enticements in favour of something grittier and, on the face of it, far more dangerous. There's a new breed of water vole in town, and they're taking over.

If we could travel back to a time before *Homo sapiens* started to engineer the British landscape for its own purposes – let's say 10,000 years ago – what would we find? There was plenty more woodland, and water voles don't care for that. Fortunately, the inconvenience would have been outweighed by substantial advantages. Except for limestone regions and hills with free-draining soils, the entire landmass would have been soggier than today – a mosaic of marshes, swamps, bogs, floodplains, streams, tributaries, lakes and ponds. All of that goes a long way towards making space for 6.8 billion water voles, but it still isn't sufficient. Maybe some of those voles lived wherever they liked in this damp environment, regardless of their proximity to water. After all, they were given their name long after human activity had narrowed their options.

Evidence in support of this theory comes from modern Europe. Across parts of its continental range, the water vole is treated as an agricultural pest, reaching implausibly high population densities on boom-and-bust cycles and causing mayhem in orchards and arable fields. If you search the scientific literature for anything about water vole conservation outside the UK, you'll be deluged with papers describing damage caused by water voles, with accompanying discussions of how best to control their populations. The water vole has a native range extending across Europe to western Asia, and it's classified by the IUCN as a species of Least Concern. No country outside Britain (maybe with the very recent exception of Italy) gives a moment's thought to water vole conservation.

The Europeans look on in wonder at our ability to bring a small and occasionally troublesome rodent to the edge of extinction. One of the many reasons why their densities are so much greater is that water voles across the Channel can be not only riparian but 'fossorial'. We need to explain our explanation because it's so utterly unhelpful: almost nobody knows what fossorial means, and those fortunate few who define it as 'burrowing' will be puzzled because all water voles burrow. Specifically, a fossorial water vole burrows in grassland; we've seen rather arbitrary definitions insisting that a water vole becomes fossorial when it lives a certain distance from the nearest watercourse, but in reality the exact distance stops mattering after just a few metres. These fossorial water voles spend more time underground than their riparian counterparts, but they're apparently the same species. The fossorial lifestyle is fairly common across Europe; in mainland Britain it seems unique to the East End of Glasgow, and the behaviour also persists on a small handful of islands in the Sound of Jura. If there are other colonies out there, we haven't found them yet.

We know from archaeological and historical records that fossorial water voles were once common across Britain. It's estimated that before humans intervened, they would have outnumbered riparians by about twenty-nine to one. That had changed long before the eighteenth century, when fossorial water voles were still putting in the occasional appearance but infrequently enough to cause puzzlement. Gilbert White wrote in 1767 that there seemed to be 'two sorts of water rat', and in December 1769 he recorded that his neighbour had uncovered one in Selbourne while ploughing 'far removed from any water'. A century later, that daunting Victorian page-turner, Llewellynn Jewitt's *Grave-mounds and their Contents* (1870), reported the discovery of gnawed human bones in a barrow on Hitter Hill in the Peak District, and went on to identify the culprit in a lengthy footnote:

> [Water vole bones] are very abundant in Derbyshire barrows, and, indeed, are so frequently found in them, that their presence in a mound is considered to be a certain indication of the presence of human

> remains . . . The part of the matter which is curious to the antiquary is, that the bones in Derbyshire barrows are frequently perceived to have been gnawed by the scalpri-form incisors of these animals.

Water voles are almost exclusively vegan, but they'll gnaw on any old bones that happen to be lying around simply as a way of maintaining and sharpening their gnashers. Barrows must therefore have been a great place to live, even if the water vole colonies had long since gone by the time that those dusty antiquarians arrived on the scene. Both White and Jewitt speculated that water voles used grassland refuges as hibernacula. What neither of them realised is that voles don't hibernate. They were living there all year round.

The Derbyshire barrows described by Jewitt provide a likely answer to a mystery 400 miles further north on the coast near Banff. In the late nineteenth century, local communities were terrified by the 'earth hounds' or 'Yird pigs' that were said to frequent graveyards. These earth hounds were blamed for 'burrowing among the dead bodies and devouring them', with one particularly Gothic account claiming that you could hear them 'crunching the coffins ere the mound was covered in'. That kind of detail has led some experts to conclude that earth hounds were nothing more than a feverish legend born of morbid Victorian imaginings, but the historical descriptions speak for themselves: the critters were half rat and half mole, with dark fur and a head like a guinea pig's. They lived underground, and several had been dug up by farmers ploughing nearby fields. A local claimed to have caught one, and demonstrated that it was nothing more scary than a water vole. But who was ever going to believe that tall tale when there was no suitable water nearby? And you might feel a bit daft if the monster terrorising your community turned out not to be some nightmarish carnivore previously unknown to science but a harmless grass-nibbling rodent.

Long before American mink arrived on this side of the Atlantic, water vole populations were already in steep decline. Land had been drained, watercourses canalised, and bankside vegetation destroyed. Fossorial water voles disappeared even sooner than their riparian counterparts. Their meadows were ploughed or overgrazed. Non-native species such as the rabbit outcompeted them, and when they weren't being deliberately targeted as crop-ravagers, water voles were mistaken for rats and killed. Forced to survive solely on waterways, they'd become easy prey if ever a supremely efficient semi-aquatic predator showed up. But what were the odds on that? Then from the 1920s mink farms started to spread across Britain.

After the Second World War, the numbers of fur farms increased rapidly, reaching a peak of around 700 in 1962. Mink escapes were reported almost from the outset. In 1956, breeding populations were confirmed on the River Teign in Devon, and by 1968 feral mink were recorded across more than half of England and Wales as well as much of lowland Scotland. The Ministry of Agriculture set up a team to tackle the mink problem in 1965, having half-heartedly tried a bounty scheme for a few years before that. The team was absurdly underfunded with no prospect of success, so in 1970 the Ministry threw its hands up in the air, withdrew the funding altogether and closed the scheme down.

Animal rights activists regularly get the blame for releasing mink into the wild, but they were latecomers to this particular disaster. Mink aren't the easiest creatures to keep contained, and sometimes their escape attempts were given a helping hand: new government regulations and formal inspections were introduced in 1962, prompting backyard enterprises to open the cage doors and count to a hundred. (We're reliably informed that this happened several times in our part of East Devon alone.) Fur was also beginning to become unfashionable, so the economic incentives no longer existed. The last twenty mink farms were closed by 2002, following a complete legislative ban on fur farming in Britain. It's miserable to discover that such a grotesque industry is still common elsewhere in Europe. In 2021, Denmark culled fifteen million mink after diagnosing them with a mutated strain of Covid. Something like four hundred other mink farms in at least eight EU countries were also infected. After culling all the mink in its sixty-eight farms, the Netherlands is now pushing for a ban across the EU.

Amidst Britain's tale of woe, there are tentative grounds for optimism. It's possible that the recovery of otters has depressed mink populations, although evidence is still sparse. Otter populations in the 1950s and '60s declined dramatically owing to the twin pressures of hunting and water pollution. They were nearly annihilated by the late 1970s, when the first National Otter Survey found that fewer than six percent of surveyed sites in England had any evidence of otters. Surveys in Wales and southern Scotland weren't much more successful. Across much of Britain, mink had the rivers to themselves.

Thanks to legislation banning otter hunting and organochlorine pesticides, otters seem to have made a steady recovery. That may yet prove short-lived: the latest research in Wales indicates the first significant decline since surveys started in 1977, and we await the results of the sixth National Otter Survey of England, run by Fiona and colleagues at the Mammal Society. Regardless of statistical wobbles, there are certainly more otters around now than there were a few decades ago when mink were sweeping across the landscape. And otters, it's fair to say, take a dim view of mink. The veterinary pathologist Vic Simpson performed hundreds of autopsies on dead otters and mink from his base in Truro, and used to horrify audiences with pictures of wounds inflicted in fights between otters; for example, ungentlemanly males will bite off each other's testicles if need or opportunity arises. He found evidence of healed facial wounds on otters that appeared to have been caused by mink, but never came across living mink with old injuries inflicted by otters. At a quarter of the weight, mink don't survive violent encounters with their fellow mustelid. So otters reduce mink numbers by killing them, but their very presence is also enough to persuade mink that they should make their excuses and skedaddle.

Early research suggested that where otters were recovering (or where they'd been reintroduced in the upper Thames), mink were declining and water voles were holding their own. Recent accounts are more guarded, not least because mink scats become difficult to detect in places where there are otters. No mink in its right mind advertises its whereabouts to a heavyweight killer. There are also regional examples that don't fit the expected patterns, such as the continued mink colonisation of some parts of Scotland despite the maintenance of good otter populations there. The quality of the data on which to infer anything at all is desperately poor, and it's impossible to deduce a trend without better information. We expect otters to reduce mink populations, but we still can't be sure.

Programmes for systematic mink control are currently operating in fewer than half of English counties, and most of these are limited to a single site such as a wildlife trust reserve, with no joined-up strategy at a scale relevant to mink behaviour and ranges. This lack of coordination is hardly surprising. Water vole project officers themselves became endangered when funding agencies, forever pursuing novelty, failed to renew contracts for work that was seen as routine or ongoing. Funding for a reintroduction scheme with lots of lovely photo opportunities and bucket-loads of public engagement? *Tick!* Funding for a year or two of mink control? *OK, if you really must.* Funding to monitor the sites for the next ten years and control mink in perpetuity? *Unfortunately, your application has not been successful.* Too much money has already been frittered away on localised culling, accompanied by some sort of small-scale water vole reintroduction project in an isolated area with unstable habitat. It's pointless and destined to fail, but it generates a few feelgood headlines en route.

This is exasperating because monitoring for mink isn't particularly hard. The animals are inquisitive and have a convenient habit of climbing onto specially designed 'mink rafts' that are equipped with a wooden tunnel and a tray of wet clay. As soon as a mink signals its presence by leaving footprints, traps can be set in the same tunnels. New advances make the process even more efficient. A 'mink police' alarm sends a text message reporting that an animal has been captured. None of this, of course, avoids the need to get consent from multiple landowners, and it still takes time to deploy, check and (when they're stolen or washed away in floods) replace the rafts. Most of all, someone proficient with firearms needs to be on hand to kill any mink that are trapped.

Thanks to human greed, mink are in the wrong place. If you have to cull them, do it once and do it right so that you're not culling them for generation after generation. The only ethical cull involves a huge boots-on-the-ground commitment; as military wisdom has it, 'Amateurs talk strategy. Professionals talk logistics.' Scotland provides a model for this hideous task: a major mink removal project that began in the Cairngorms has now expanded to cover about a third of the country. Based on excellent scientific research by Xavier Lambin's team at the University of Aberdeen, the project seems to be making genuine progress even though the impact of culling is partly offset by better breeding success.[1] Mink were successfully eradicated across the original target area of 10,570 square kilometres. The Scottish Mink Initiative has now expanded to cover 29,000 square kilometres, and, although funding remains an ever-present issue, the scheme has sustained mink control for longer, and over a greater area, than any other in Europe.

It's undeniably helped by the fact that much of the mountainous region in Scotland doesn't offer great habitat for mink. By starting in the upland headwaters, the researchers were able to create refuges for water voles while pursuing mink control further downstream, where occasional incursions are more likely. Another bonus (at least for this project) is that so many locals are directly or indirectly involved with the nearby shooting estates. The result is a greater acceptance of the need to kill wild animals and a local antipathy towards any invasive species that aren't game birds – and there's no shortage of people with the relevant gun licences. In Scotland (unlike, say, Wales or Devon), significant chunks of land are owned by a relatively small number of people, so support is easier to come by. Then there's the helping hand provided by Scotland's enviable Land Reform Act (2003), which allows volunteers the freedom to wander unhindered along riversides.

English efforts are more lacklustre. The only significant attempt at wide-scale mink control is underway in East Anglia, with backing from the Heritage Lottery Fund. Riparian water voles are doing relatively well in the flatlands; they have a network of ditches, fens and slow-flowing rivers to splash around in, and densities can reach higher levels in these two-dimensional networks than in the linear waterway habitats that dominate most of the rest of England. That's also the reason why East Anglia is uniquely challenging when it comes to mink control. Very often in Scotland or Wales, mountains act as a break on expansion, and in areas like the Midlands urbanisation can halt the movement of mink across a landscape. East Anglia has none of those advantages, and the homogeneity of its landscapes and agricultural practices means that it's difficult to transfer any lessons learned into regions with more variable topography. As many experts pointed out before it carried on regardless, Natural England's decision to locate its national pilot study of water vole recovery here seems like an opportunity wasted.

We'd like to volunteer some frivolous facts involving water voles.

A water vole needs to eat eighty percent of its body weight every day. That increases to 200% if it's a lactating female. It can give birth to up to five litters every year, with a gestation period of twenty-two days and an average of five or six pups in each litter. Like many prey species, it lives fast and dies young. Few water voles get to celebrate their first birthday.

Rob Strachan worked out that water voles operate on a four-hour cycle: one hour above ground for every three below. That changes according to season, and presumably fossorial voles spend less time out of their burrows because they're more exposed.

You'll sometimes hear it claimed that water voles are good swimmers. That's a charitable interpretation. They don't have webbed feet, their fur quickly gets waterlogged, and they can only dive for twenty seconds. Their special trick of kicking up sediment to evade predators doesn't work at all against mink.

Males are less aggressively territorial than females. We've watched riparian females wrestling each other, indifferent to our proximity. This behaviour drives juvenile females away into new areas. Everyone calms down when population densities are high: if you're having to fight all the time, you're extremely vulnerable to injury and predation. Fossorial water voles seem to get on better, perhaps because they can extend across a 360° range. They're not imprisoned along a narrow strip like the riparians.

Read's Island in the Humber Estuary was the scene in 1896 of what one local newspaper called a 'plague of water rats'. The first remedy was to flood its 500 acres at spring tide, but that proved only partially successful. A shooting party then spent a day blasting anything that moved, still to no avail. Finally, nature stepped in, and the same cycles that caused the surge in numbers brought about the inevitable collapse. Today Read's Island is an RSPB reserve, and home to ten percent of the UK's population of avocets – memorably described on the RSPB website as 'birds with bizarre upturned beaks'.

The first house we ever owned was a two-up two-down end-of-terrace railway cottage in Chippenham. The reason that we could afford it lay twenty metres away, on top of the embankment that loomed over our front door: the Paddington to Bristol train line. We loved that house. The trains proved far less noisy than road traffic, and the embankment itself was a corridor for all kinds of wildlife. Here we made an independent discovery that got reported by a real scientific team conducting the same experiment a few years later: if you Tippex numbers onto the shells of snails that are eating your plants, then translocate (i.e. hurl) them a great distance into the embankment's greenery, you'll find that most will slither back again a few days later. Not that we minded at all: we had our own little wildlife oasis in the heart of a busy town. But as Robert Frost once warned, 'Nothing gold can stay'. Our home was just the right size for us. It was the wrong size for us and fifty rodent house-guests. They weren't gold, and they were staying.

At the time, Fiona was working for the Wildlife and Conservation Research Unit at Oxford. WildCRU was in the vanguard of conservation groups sounding the alarm over water voles. Rob Strachan had spent two years (from January 1989 to December 1990) travelling all over Britain in a camper van to survey for water vole presence. He found that their numbers weren't so much declining gently as collapsing catastrophically.[2] When British Waterways wanted to carry out some repair work on a stretch of the Kennet & Avon canal between Hilperton and Seend, they called in WildCRU to round up all the water voles and pack them off for an extended holiday in pre-prepared lodgings. Meanwhile, the contractors were supposed to fix the leaky canal, stabilise the banks, and recreate the water vole habitat according to strict specifications.

On the morning appointed for the release, Fiona and her colleagues found that the site had been made utterly uninhabitable. Any ecologist who advises developers will recognise this all-too-familiar kind of snafu: against clear instructions, the contractors had added the thinnest layer of soil – we're talking inches – over concrete and metal.

The water voles couldn't go back to their temporary accommodation because it was no longer available. Their options were to be dumped on the banks and die very quickly from lack of shelter, or come and stay with us while someone tried to botch together some sort of solution. That's how we ended up living for months with fifty water voles, each in its own cage, and each very fragrant. We even had to pay for the hay, apples and carrots ourselves. Were the water voles grateful? Based on their willingness to sink their teeth into any human hand trying to clean them out, we're guessing not. We managed somehow to keep all but one of them alive – a far lower rate of attrition than they'd have experienced in the wild – but there was no happy ending. Eventually the contractors added coir rolls to the banks, in the hope that vegetation would grow and that the water voles could burrow down through it after release. Planted coir can provide good water vole habitat,[3] but in our case it was being laid on top of concrete, and the vegetation only had weeks to get established before we were out of time. It was a choice of releasing before autumn set in or waiting until spring, by which point we'd have a herd of habituated voles and fewer fingers. And so we bade a worried farewell to our furry lodgers. When the site was surveyed by WildCRU a couple of years later, they'd disappeared.

*

What that fiasco demonstrated is that culling mink is necessary but not sufficient. We shouldn't even be thinking about reintroductions without also seeking to reverse the damage inflicted on wetland habitats over the past few centuries. Around 6.4 million hectares of agricultural land have been drained with piped systems in England and Wales alone; in the process, ponds that provide refuges from mink have all but vanished.[4] It's not just water voles that have gone with them, but amphibians, invertebrates and wetland plants. Not everyone cares: the government's Agriculture and Horticulture Development Board produces a booklet for farmers about land drainage that somehow fails to muster a single mention of the damage to biodiversity. Overgrazing and the compaction of farmland by heavy agricultural machinery, together with the tarmacking of new roads and other hard surfaces, ensure that rivers are increasingly prone to sudden periods of spate that destroy bankside vegetation and drown water voles in their burrows. Water voles like to paddle across slow-flowing streams and mosey around in the cover of tall vegetation; unhelpfully, humans reduce rivers to fast-flowing conduits.

Given the scarce funds and the fragmented populations, it makes sense to prioritise areas for water vole conservation that are likely to deliver the biggest wins. Fiona and her Mammal Society colleagues have been trying to identify those areas for Natural Resources Wales as well as for the East Anglian water vole scheme. One obstacle is that decent data on water vole population sizes and distributions don't exist; it's easier to find information about the surface of the moon than to discover whether banks are poached by cattle, how deep the waterside vegetation is, or the extent of any mink control. Applying the principle that all models are wrong but some are useful, the Mammal Society's maps nevertheless provide tentative suggestions for where habitat improvements and vole releases will be most effective.

At least they've put a stop in Wales to one action that would directly damage water vole conservation – tree planting on riverbanks. Water voles eat non-woody vegetation, but won't stray far from home to get it. Breeding females often refuse to budge more than a body length from their burrow. The greater the availability of forage, the faster the juveniles reach sexual maturity, and the smaller their home ranges need to be. The result is a higher population density of water voles.[5] Not surprisingly, reintroduction programmes are more successful along banks with deep vegetation of three to six metres than along narrower strips.[6] Tall-growing trees provide a serious problem, because the shade thrown across the bank limits plant growth, which in turn reduces food supplies and leaves voles vulnerable to predators. Some conservationists even speculate that the First World War contributed to the decline of water voles because trees that would previously have been coppiced or pollarded were left to grow. (For similar reasons, the disappearance of beavers has also damaged water vole numbers.) Nowadays, difficult-to-farm areas such as river corridors are attractive places to plant trees, so the potential for conflict between forestry subsidies and water vole conservation is high. Natural Resources Wales is well ahead of other home nations when it comes to planning how to achieve targets for carbon-sinking woodlands; it has identified Woodland

Planting Opportunity Areas that specifically exclude water vole strongholds.

So much for the riparians. As for fossorial water voles, if we want to answer the question, 'Why Glasgow?', we need to bear in mind a tantalising possibility: 'Why not everywhere?' Normally, we think of the countryside as a last sanctuary for wildlife, protected from the ever-intensifying pressures of development. Now the roles are reversed: maybe urban rewilding can save the water vole from extinction in Britain. Research in Glasgow has discovered one extraordinary fact: water vole densities are highest in areas with the greatest human disturbance.[7] That's lucky because we're experts at human disturbance. Towns and cities offer uncultivated grasslands free from livestock and ploughing, and rabbit populations have shrunk owing to myxomatosis and viral haemorrhagic disease. Best of all, people provide a buffer against mink, which won't stray far from water and certainly won't venture into heavily built-up environments. What Glasgow has today, every one of our cities could have tomorrow: a profusion of water voles. Take away the farmer, the competitor and the predator, and suddenly it's happy days for the little scamps. All we're lacking is the vision thing.

While we wait for someone to carry out the genetic analysis, there are three theories about the origins of Glasgow's fossorial water voles: they've always been there, someone has deliberately released them, or they've migrated from neighbouring areas. This last is the most frequently cited as well as the most appealing. The story starts in the late 1960s with the building of the M8 motorway that joins Glasgow and Edinburgh. Urban planners decided in their wisdom that the motorway should run straight through the middle of Glasgow. To bring about this utopian masterstroke, they needed to overcome an annoying inconvenience: there were quite a few buildings in the way. The obvious solution was to follow the course of the Monkland Canal, going right over the top of it. Neither the politicians nor the developers cared about the wildlife; water voles had no legal protection at the time. As for the voles themselves, they didn't appreciate the intervention, so they dispersed into the surrounding fields, where they soon adapted to their new circumstances by becoming fossorial, and lived happily ever after while their riparian brethren got eaten by mink.

This account has other things to recommend it besides its neatness, the most compelling being that all the fossorial water vole sites occur within a few hundred metres of the motorway. Even so, it does require one substantial leap of faith. It wasn't until 2008 that the first colony of fossorial water voles was identified in Glasgow. Pest control officers were called out to deal with a rat problem and, thankfully, one of them noticed the difference. After that, it was open season: water voles started popping up all over the place. Officially, there are now more than sixty occupied sites in the East End of Glasgow, and on our last trip we found and reported a couple more. So we're supposed to believe that, for almost forty years, water voles had lived alongside humans in a tightly packed urban environment, and not a single person had noticed or thought to comment. Rat-catchers, dog-walkers, park-keepers, street-sweepers, environmental consultants assessing sites for possible development:

not a soul. Of course, they had every reason to expect rats rather than voles, no matter that water voles were dancing a jig down the high street. This sort of behaviour seems to have ended badly for one poor individual that we spotted on our last visit. It was laid out dead on full display, next to a zebra crossing not far from a Greggs bakery. And it definitely wasn't a rat.

In its favour, the motorway origin theory for Glasgow's fossorial water voles does have the virtue of being the least unlikely of all possible options. If we wanted to argue for deliberate translocation instead, the obvious source would be one of those very small islands in the Sound of Jura, a six-hour drive west of Glasgow. A census in the 1990s showed that there were plenty of individuals to spare: their density in spring reached 2,600 per square kilometre, compared with the measly five per square kilometre that you'd expect in riparian colonies on the mainland. Why, though, would anyone have thought that it was a good idea to catch some water voles, come back to the mainland on a boat, make the long journey to Glasgow and drop them in a municipal park in the city's East End? People do the weirdest things, but no matter how amazing the outcome, even the most optimistic rewilder wouldn't have foreseen what was going to happen next.

Here's another problem. We know from DNA sampling that water voles came to Britain in two distinct waves.[8] The first of these originated in the Iberian Peninsula, and at one point would have colonised the entirety of Britain. A second wave from Eastern Europe turned up later, probably via the land bridge before it was inundated over 8,000 years ago. These pointy-elbowed arrivistes took over, replacing the original water vole population throughout England and leaving them confined to their Scottish stronghold: the division between the two lineages maps fairly neatly onto the modern border. Scottish and English water voles don't generally mix, although a few of the Scottish sort have lingered in Northumberland. The Scots are more likely to have black fur and the English brown, yet you'd struggle to notice any variance in size. One study found that the average Scottish water vole weighed in at 191 to 205 grams, and the English water vole 196 to 219 grams, so the chunkiest English water vole was chunkier than the chunkiest Scottish water vole by the equivalent of a tablespoon of butter. A mature male water vole can actually reach 300 grams or more, with the English still tending to be fractionally bigger than the Scottish.

Turn to Glasgow's fossorial population and we can instantly see the difference. As part of her doctoral research, Robyn Stewart measured over 100 Glaswegian water voles, calculating their average weight as 109 grams. She was reluctant to draw general conclusions owing to sample size and the fact that a high proportion were juveniles, but fossorial water voles in Europe also tend to be smaller than their riparian counterparts. Our own experience, albeit with an even smaller sample, is that the water voles in Glasgow are noticeably smaller. They weigh about four times more than a bank vole or field vole, but they're half the size of those exiled riparian water voles from which they're supposedly descended. (Conventional wisdom has it that a water vole needs to be at least 170 grams to stand a decent chance of surviving the winter.)[9] We have to believe that in the course of, say, thirty generations, they did an awful lot of shrinking. That could only happen if there were massive selection pressures that exclusively affect fossorial populations. Do hefty water voles get eaten in grassland because they're slower or they need bigger burrows? Are fossorial water voles smaller because their colonies are so much more – sometimes dozens of times more – densely populated? Or is this evidence that we're considering two different subspecies?

Deep breath. Taxonomy. Despite the obvious differences in their lifestyle, the current assumption is that fossorial and riparian water voles are the same species. This orthodoxy could change and change back again by this time next week; taxonomists like nothing better than splitting species and then lumping them together into the same or different groups. Even reputable organisations like the Global Biodiversity Information Facility (GBIF) and the IUCN haven't managed to keep up with the taxonomical hokey-cokey.

The first thing to get straight – and this won't be overturned – is that water voles aren't rats. It's never a great PR strategy to be confused with the most hated creature in the animal kingdom, but the problem has plagued them since at least the seventeenth century and no doubt long before. In 1693 the naturalist John Ray gave them the name *Mus major aquaticus* – greater water mouse, or water rat. It took another hundred years before a French zoologist rejoicing in the name of Bernard-Germain-Étienne de la Ville-sur-Illon, Comte de Lacépède, worked out that they were in fact voles. As in modern-day Glasgow, the delay was understandable up to a point: habitat and fur colour are similar, although a more detailed inspection would easily discriminate on the basis that rats have pointier muzzles, more prominent ears, and tails that are long and scaly. Water voles are blunt-nosed, which is a diagnostically volish and not at all a ratty characteristic. Even so, you'll still encounter abominations like this nature-loathing definition of 'vole' provided for children by the Merriam-Webster online dictionary: 'a small animal that is a rodent which looks like a fat mouse or rat and is sometimes harmful to crops'. (There's so much wrong here, not least the clunky expression, that we don't know where to start.) Just as we still sometimes call them water rats, many European languages refer to water voles as mice or rats. In German, for example, they're *Wasserwühlmaus*, which is literally as it sounds: 'water vole mouse'.

Carl Linnaeus, Sweden's most famous export before ABBA and IKEA, arrived in the eighteenth century to sort out all this nonsense. A botanist by training, Linnaeus invented the modern system of taxonomy by which organisms are assigned, with increasing levels of precision, into non-overlapping groups labelled Kingdom, Phylum, Class, Order, Family, Genus and Species. (After years of getting the sequence muddled, Fiona finally found the mnemonic guaranteed to sear it into the brain: Keep Penis Clean Or Face Getting Syphilis.) Linnaeus based his classification on the blindingly-obvious-with-hindsight principle that organisms in the same group would share characteristics distinguishing them from organisms in different taxonomic groups. There were limits to his system, of course. He was working 100 years before Darwin (who went on to add the idea of separate species being unable to interbreed successfully) and Mendel (who founded modern genetics). Nevertheless,

Linnaean classification still stands up remarkably well. Common and vernacular names don't travel reliably across linguistic and geographical borders, but conservationists across the world can refer to an organism by its two-part Latin name – genus then species – and be sure that they're talking about the same thing.

Linnaeus was evidently perplexed by water voles. His treatise *Systema Naturae* (1758–59) describes two species on the same page: *Arvicola amphibius* and *Arvicola terrestris*. These names seem to separate riparian and fossorial species. Etymologically, *arvicola* means 'inhabitant of a cultivated field', but *amphibius* points to a semi-watery lifestyle while *terrestris* emphasises land. Although arguments have raged over the centuries, since 1970 most taxonomists have accepted that *A. amphibius* and *A. terrestris* are the same species. There's also a separate European species, *Arvicola sapidus*, the southwestern water vole, with a different number of chromosomes; it's restricted to the Iberian Peninsula and France, and was once a regular ingredient in paella. (We're vegetarian, but even if we weren't . . .) When it came to the British brand of water vole, for a long time *terrestris* and *amphibius* appeared interchangeably as species names, with the former more common in scientific research. Finally, someone pointed out that the first recorded use of the word *amphibius* predated that of *terrestris*, so that became the official name.

In 2001, a Russian scientist examined multiple specimens of fossorial and aquatic water voles and concluded that the former could be distinguished as a separate species, *Arvicola scherman* (the montane water vole). It was described as smaller, with softer fur, fewer bumps on the soles of its feet, and incisors that projected more strongly forward. This new classification was widely accepted – but there's a problem. When a species splits in two, we expect to see a single branching point in the family tree (known as a monophyletic origin): one line of the water vole family should have gone off to gambol through the pastures, while the other grabbed its towel and goggles and headed for the nearest watercourse. Over time, the voles in the different environments would encounter each other less frequently, experience different selection pressures (such as the need for different tooth angles and more enamel in the burrowing form), and undergo different random mutations. What had been one species would eventually become two. Very neat, but that's not remotely how the family tree looks. Genetic analyses identify clear differences between water voles from different geographical areas, with separate lineages in Turkey, Italy, western Europe and the rest of Eurasia. (The Italian voles, for example, look like they're on the verge of forming a separate species.) Yet when we compare characteristics of *A. amphibius* and the putative fossorial water vole *A. scherman*, we see branching points occurring separately in the different geographical locations. In other words, rather than the terrestrial and aquatic forms dividing and then colonising the different regions of Europe, fossorial and aquatic lifestyles seem to pop up independently in each region, suggesting a high degree of natural flexibility in behaviour and morphology.[10]

We know from our own experience that when you put aquatic water voles in a pen without water, they'll start digging. Derek Gow, breeder of more than 25,000 captive water voles for reintroduction schemes around Britain, also reports that water voles from aquatic environments adopt fossorial habits if they escape into a field. Maybe all aquatic forms have fossorial behaviours stashed away, ready to use in an emergency. Given enough time apart, terrestrial and aquatic populations should separate into different species, but where our British voles currently sit on that spectrum is anybody's guess. One study managed to get the two forms to breed successfully; another failed. Some publications are hedging their bets and describing the fossorial form as a subspecies, *Arvicola amphibius scherman.* The European Environment Agency website lists *Arvicola scherman* as a species but reports that it 'has very limited information', which is a circumspect way of saying that it has absolutely no idea what's going on. And if you're wondering why we're chuntering about these rather niche taxonomic debates, it's because for once they do actually matter. Let's say that our Glaswegian water voles can switch freely between terrestrial and watery habitats, as is sometimes reported, and that they're the same species as the riparians: these colonies could be the salvation for the species all across Britain. Why waste time and resources clearing out mink to establish precarious riparian populations when you can just dump a few water voles in the middle of a field? But if they're well on their way to becoming a different subspecies, with their own adaptations for life away from water, then we need to pursue separate conservation efforts for the fossorial and the riparian forms.

Appropriately, Cranhill Park was our first stop in Glasgow: a large metal structure by the entrance announces that 'Water voles were discovered here in 2008'. We admit to having been a little apprehensive. As Wikipedia records, Cranhill was once 'infamous for its illegal drug trade and antisocial youth culture', and for a time it became known in the media as 'Smack City'. A lot has changed for the better in recent years, but the site still presents unusual problems. A previous survey of water voles had come to a sudden halt when thirty traps were stolen; we've no idea what the exchange rate between hard drugs and water vole traps happens to be, but there must be a dealer somewhere with a lock-up full of them. Another survey had been halted owing to what the subsequent scientific thesis tersely reported as 'verbal threats'. We were advised not to trap there at night, and not to leave any equipment unattended for more than a few minutes. A pair of nerdy English environmentalists gallivanting around the mean streets of Glasgow: we felt like red meat. We soon relaxed when every interaction with the locals turned out to be positive. They were keen to find out what we were doing, and delighted that their water voles were getting attention. One elderly chap on a mobility scooter stopped us with the best opening gambit we'd ever heard: 'Are you the water vole people?' Yes, yes we were. He pondered for a moment. 'I saw Jeremy Clarkson on TV, and he wasn't allowed to dig out a ditch on his farm because of water voles. He said, "I'd heard they were all in the East End of Glasgow." I thought he was joking.'

We'd parked on the road outside Cranhill Community Centre, which had been requisitioned as a Covid testing site. The first thing we saw, right by the fence adjoining the park, was a heron. We knew exactly why it was skulking round an inner-city location so far from water. Derek Gow tells how his first release at Barn Elms in London ended with a heron effortlessly spearing the newly reintroduced water voles. He goes on to conclude, against his initial judgement, that this equals 'success'; water voles are meant to be eaten by native species, or at least some of them are. Still, we're not going to lie: we find the heron a difficult bird to love. Later the same day, we came to a patch of the parkland that had historic signs of water voles but no evidence of recent activity. A passing walker said that we wouldn't find water voles there because the herons had cleared them out. They swallow them whole, he assured us.

Robyn Stewart had surveyed Cranhill twice before, coming up with estimates of seventy-eight and forty-two individuals. That second number probably didn't represent a real decline: water vole populations fluctuate wildly according to season and, with an average longevity of just a few months, most won't survive a cold spell. We were there in August, which is probably peak vole because there's plenty of food and cover and a high proportion of juveniles. It wasn't long before we discovered an area that looked promising, a dozen metres away from the tarmac paths. There, among dog poo and crisp packets, were the burrow entrances, latrines dotted with what looked like black Tic Tacs, and signs of recently chewed stems. Several of the holes had been clogged with tiny bundles of grass. This is a multi-use behaviour to keep out rain, provide food for later without the need to leave the burrow, and camouflage the hole. Tim stood over one entrance, wondering aloud whether it would be a good place to set a trap, when out of a thicket tumbled a water vole, darting down that very hole like black lightning. This was going to be easy. After that, of course, we didn't catch a single vole all morning.

Our mood as we ate our limp sandwiches was downcast. We'd arrived with such hopes. Now the rain was moving in, not heavy but steady. There were still a few people wandering through the park, sometimes accompanied by dogs, but we didn't believe that anyone would see, let alone steal, the traps in those conditions. Fortified by hot tea, we trudged dutifully out to check them again, and there were our voles waiting for us. In four days of trapping in changeable weather, we only ever caught voles during wind and rain. So began a theory that we hereby offer to the world as the Kenthews Hypothesis: the wetter the conditions, the easier it is to catch water voles. Who needs fiddly things like statistical significance and confidence intervals when you can make do with cherry-picking and anecdote instead?

A burn runs below the park at Cranhill, and it's visible for part of its journey. The water voles didn't seem especially interested in those damp sections, although again there were signs of historic activity. Perhaps the herons had finished off that colony, too. Robyn Stewart has noted that in Cranhill the fossorial voles seem able to switch between wetland and grassland habitat, although we don't know how or why. Anyway, we tried our traps by the burn, but no luck. We had more success further off, and by the evening we'd caught a grand total of four voles: two black and two brown. (The fossorials don't seem to be so strictly segregated by colour as the riparians.) Fantastic though that was, it didn't prepare us for what we'd find the following day.

You take your square guttering and cut fifty-centimetre sections with the handsaw. You add chunks of apple, leave it in the right place, and wait for a water vole to visit. By all accounts, Pink

Lady is the apple of choice for the discerning water vole. It'll pause to fill its chops with the exotic treat and, if you're lucky, will return the favour by leaving behind a poo sample before scuttling on its way. Everybody wins. If you need the whole animal, take your specially designed cage trap with a treadle mechanism. You put the apple on the far side of the treadle. The water vole, intent on the apple, runs in, triggers the mechanism on its way to the feast, stuffs the food, and retreats into an attached wooden box that's been filled with hay. There it hunkers down and waits to see what'll happen next.

A short time later you check the trap. Noticing that the door has been closed, you take the cage and empty its contents – hay, apple, water vole all helter-skelter – into a large polythene bag. You now have in the bottom of a deep bag a very cross mammal that faces you with spectacular orange incisors at the ready. Even though you're wearing gloves, you didn't get up that morning intending to be bitten. This is where the Pringles tube comes in handy. You hold the tube in the bottom of the bag and the water vole runs inside, head first, on the principle that if it can't see you, you can't see it. You lift out the tube, hold the vole by the root of the tail (or, if you can reach, the scruff of the neck), and pull it out so that its front paws are now resting on the tube's rim. At that point, you can determine its sex, take a mouth swab, check for parasites and so on. When the water vole has done its bit for science, having traded the joy of the apple chunk against the stress of being handled, it's released back into the grass, where it disappears faster than a wig in a storm.

When we visited Glasgow in 2021, none of the above required a licence; Scottish law had been out of step with

English and Welsh since 2008. As NatureScot disapprovingly explains, 'legal protection is currently restricted to the water vole's places of shelter or protection and doesn't extend to the animal itself'. We'd only have needed a licence if we wanted to bury a trap within a water vole's burrow. It's gobsmacking to think that one of the most important wildlife populations anywhere in the British Isles should for so long have been left exposed to disturbance and destruction. NatureScot is now openly advocating for protection to be extended in law to include the water vole as well as its habitat, which should take place as soon as bureaucracy allows – that's to say, with glacial urgency.

The current situation does, at least, offer Glasgow City Council some cover when it's criticised for leaving the place looking 'messy'. It's not really a stretch to define a water vole's 'place of protection' as including the habitat around the burrow. Grabbing a lawnmower and hacking everything to oblivion isn't good for voles or for anything else, but developers and local councils across Britain regularly send in the strimmers. You can see why they do it, because despite the extra costs, it gives them a quiet life. By taking a principled stand, Glasgow City Council drives a small number of residents into apoplexies, apparently on the basis that – as one of them perceptively pointed out – water voles don't pay council tax. Social media is where common sense goes to die, but even online the vast majority of Glaswegians express pride in their water voles and a desire to find out more about them. Our own conversations overwhelmingly backed up that impression. One resident of a tower block told us how relieved she was that the next field was home to a colony of water voles, because she knew that otherwise it would be concreted over and developed like everywhere else, and she enjoyed looking out onto what she called her 'patch of green'. Conservation laws, she told us, don't just benefit wildlife.

We keep our own small plots of land around our house as wild as possible, and we've overheard people unfavourably comparing the state of our front garden with our neighbours' more interventionist approach. They see loveliness where we see a desert, and where we see vibrancy they see shameful neglect. We have slow-worms, grass snakes, newts, dragonflies, and mice and voles of various kinds; they have a neat lawn. But a lawn isn't just a lawn: it's shorthand for a network of social assumptions that are too deeply engrained to be easily challenged.

We dread to think what Fiona's grandma would have made of it all. For much of her life, she rented a council house in Preston. It had a large garden, a trim privet hedge, a tightly mown lawn and lots of pert roses. Having been widowed and left with four young children shortly after the war, she was grateful for the new cradle-to-grave social contract, and loudly supported Labour for the rest of her life. At the same time, she had a strong work ethic, was intolerant of people she called scroungers and layabouts, and would bemoan the community's falling standards. In the old days, doorsteps had been scrubbed, paths swept and lawns mown. There was pride in the neighbourhood. Now the lawns were growing wild, and people didn't even scrub inside their houses, let alone outside. Her view of what we might call rewilded grass couldn't have been more different from a present-day conservationist's. These are often generational and class-based divides, and the only way to break through the barriers is to celebrate wildlife via community campaigns like 'Living Churchyards' and 'No Mow May'.

At the risk of starting a stramash, we can confidently assert that long grass should figure well down the list of things for the East End of Glasgow to worry about. We've never seen so much litter as we did during our last trip. It was everywhere: chocolate wrappers, beer cans, discarded shoes, old television sets and, most commonly of all, broken glass. We're not arrogant enough to turn up in another country and point the finger – for a start we wouldn't know where to point – but someone, somewhere, is badly letting down the community. The danger is that the grass and the litter are getting conflated as twin halves of the same problem. Pick up the litter and let the grass grow!

A mile away, Sandaig Park is much smaller than Cranhill, with dimensions of about 150 by 100 metres. It's bounded to the north and the south by roads, and on its eastern and western borders lie, respectively, a primary school and a Jewish cemetery. The park is exactly the kind of place that people walk *through*, not *to*. It's almost treeless, and criss-crossed by tarmac paths; in its midst – scattered higgledy-piggledy – are a children's play area, an outdoor gym and an all-weather sports court. It has to be all-weather. We arrived in driving rain, and a cold fog sat over the park that long August day. We checked several times that we were in the right place, because it was, to be honest, the most unprepossessing of sites. Nothing about it had prepared us for the fact that it was going to provide one of the best wildlife experiences of our lives.

There were only three areas in the park where the grass wasn't mown: on its northern bank separated from a bus shelter by some railings; in a strip along the eastern side by the primary school; and towards the middle in a small patch between the outdoor gym and what Google Maps calls the Kids Playzone. No doubt the voles would spread across the entire park if there were a more sympathetic grassland management regime in place. They were first discovered here around 2016, and as the council's rather brilliant environment team told us, 'they seem particularly bold in this location'. The patch by the playground had what looked (and smelled) suspiciously like a fox earth in the middle of it, so we weren't entirely surprised that we caught no voles there; one local stopped us to say that he'd spotted a fox running towards the cemetery with a vole in its jaws a few days previously. The other two areas, though, were wonderful for trapping and, had we not been soaked and shivering, we'd have stayed for longer. The raised bank on the park's northern boundary had an awful lot of junk strewn across it. It also had an awful lot of water voles. Scatter some slices of apple and stare at them for long enough, and sooner or later a vole will pop up next to an Irn-Bru bottle. Even before we started trapping, we saw a couple of voles running around. Another spent a minute peering out of its burrow, looking straight at us and considering its options. It waited until we'd fished our camera out of the rucksack and then, with exquisite timing, retreated underground.

We caught seven water voles at Sandaig Park, six of which were black. The brown ones are undeniably cute, but the black are drop-dead gorgeous. Tim's video clip of a black water vole at Sandaig is his most-watched to date, with 30,000 Twitter viewings. That may not be much compared with the audience for Kanye's latest prognostications, but in an alternative reality far superior to this vale of tears Kanye would start retweeting cute water vole clips, and Tim's would get millions of hits. Admittedly, the voles weren't all in the best of health. A couple had a mite infestation around one eye, which is only to be expected given their extreme population density and the weirdness of their environment. It's a treacherous landscape in other ways, too. If a fox doesn't get them, a dog or cat might. A woman told us cheerfully that her dog is forever chasing the voles in the park. The fact that they're able to thrive despite all these deadly pressures brings home the extent of the damage done to riparian voles by mink and habitat loss.

Days three and four were an anticlimax. It had stopped raining, and we went to the places where, conventionally, you'd expect to find water voles. Robroyston Park was wonderful. It was also home to a heronry, and, although there were more historic signs of water voles, we couldn't find any recent activity. There wasn't a single blade of chewed grass, never mind a latrine. Our final site was promisingly grungy, but again no luck. Fiona noticed wasps milling around a nest they'd made in an old water vole hole and didn't bother saying anything; five minutes later, Tim walked straight over it and got stung twice. The atmosphere was tense for a while after that.

We were also finding evidence of active vole colonies on plenty of sites where we weren't allowed to trap, including the edges of private fields, some of which were marked for imminent building work. Driving from one site to another seemed to take hours as we stopped at every roadside verge. A quick glance at the grass and a nod of the head and we'd be back on our way to the next patch of scrub. We'd already concluded that the East End of Glasgow is the water vole capital of Britain, but the scale of the phenomenon still isn't recognised. Glasgow is a city in the midst of a massive wave of development. We urgently need to direct as much research, resource and legal protection to these East End sites as we can possibly manage. A map has recently been published online, based on work by the University of Glasgow in collaboration with the City Council and Robyn Stewart, showing the key water vole locations and the green connections between them. Developers aren't in any position to plead ignorance.

If only we'd caught our water voles on days three and four: that would have made a much more appealing narrative arc. Luckily, we needn't end the chapter in disappointment, because there's one exciting fact still to report. Anyone who's read books and essays about water voles will know we have a statutory obligation to point out that Ratty in *The Wind in the Willows* is actually a water vole. You're welcome.

FIVE

Hanging Out with Greater Horseshoe Bats

'What Is It Like to Be a Bat?' The title of Thomas Nagel's 1974 essay has provided modern philosophy with one of its most famous questions. You can tell why Nagel chose a bat: it's a placental mammal like us, but it's nocturnal, it echolocates, it hibernates and it flies. As Tim and Fiona readily attest, sometimes it can be a struggle to imagine what it's like to be your own spouse – let alone a stranger or a pet dog or a chimpanzee. From a human perspective, bats occupy the far edge of our empathy range, just before the point where creatures become so dissimilar that we've no idea what's going on in their minds, or even whether they have minds at all.

The conservationist's answer to Nagel's question is another question: which species do you mean? All bats belong to the order Chiroptera, from the Greek *kheir* ('hand') and *pteron* ('wing'). There are more than 1,400 species of bat, and they make up twenty percent of all mammal species worldwide. Our order is Primates – not quite so bountiful but still home to over 500 species. Hanging under the eaves of a bungalow in Barking is a particularly philosophical pipistrelle that spends its days trying to figure out what it's like to be a primate, all the while refusing to draw any distinction between humans, spider monkeys, orangutans and lemurs. Well, they're all pretty much the same, aren't they?

Kitti's hog-nosed bat is an insectivore that measures three centimetres and weighs two grams; the giant golden-crowned flying fox is a fruit-eater 700 times heavier. There are bats that eat nectar, bats that eat frogs, bats that eat fish, bats that eat scorpions and bats that drink blood and regurgitate it for their friends back at the roost. There are bats in northern Scandinavia, bats in the Amazon rainforest, and bats in deserts in Israel and Mexico. There are bats with tongues longer than their bodies, and bats with tails longer than their bodies. There are bats that cut up leaves into little tents so that they can hide under them. There are white bats, black bats, brown bats, yellow bats, orange bats, red bats, pied bats, hairless bats and bats with mohawks. There are bats with silly names: bulldog bat, visored bat, hammer-headed bat, greater false vampire bat, Buettikofer's epauletted fruit bat and, our favourite, the tailed tailless bat. Bats pollinate banana, cocoa, mango, agave (no tequila without it!), and at least another 500 species of plant; more than eighty medicines are derived from bat-pollinated flowers. Despite what you may have learned in school, the fastest mammal in the world is a bat, not a cheetah, and the race isn't even close. The only two mammal species exhibiting natural male lactation are bats. Nineteen species of mammal are longer-lived in proportion to body size than humans; eighteen of them are bats (the other being, randomly enough, the naked mole-rat). And to finesse our earlier claim: not all bats are nocturnal, not all bats echolocate, and not all bats hibernate – but they do all fly, and in fact they're the only mammals capable of true flight. We're sure that Thomas Nagel wasn't meaning to insult them, but let's ignore our instinctive bias and concede that bats are a whole lot more interesting and diverse than primates.

Britain is home to eighteen bat species, although in effect it's seventeen because – as we've mentioned previously – we're down to our very last greater mouse-eared bat. Although many websites insist that it's been extinct in Britain since 1990, IUCN rules dictate that a species isn't officially extinct until exhaustive searches show that not even one remains in the wild. And one definitely does remain. The Lonesome George of this particular story hibernates most years in a tunnel in West Sussex. He skipped a couple of winters from 2019 and we all feared the worst, but he was back again in 2021, hanging around his old haunt. The exact location is top-secret; if we told you, we'd have to kill you. We've no idea where he goes in summer. It's possible, but unlikely, that he darts across the Channel and spends the warmer months cavorting with flighty French mademoiselles before returning to England to recuperate. The south coast of England has always been at the northernmost edge of this species' range, but climate change means that its distribution is inching our way. That's one reason why Fiona's currently hatching a long-term plan to bolster (or re-establish) the British population with imports from the continent. It's not as if they disappeared from Britain centuries ago, like the lynx or the wolf; there was a roost with at least thirty of them a few decades back. Faced with a species classified as 'Critically Endangered' and on the verge of imminent extinction, the government hasn't yet roused itself out of indifference, but we live in hope that when the time comes, official agencies will spring into action and offer their fervent support. [STOP PRESS! In January 2023, as we're going to print, a second greater mouse-eared has turned up in Sussex.]

The greater mouse-eared belongs to the *Myotis* group of bats, along with six other bat species found in Britain. *Myotis* does indeed mean 'mouse-eared', as befits the widespread erroneous assumption that bats are flying mice: a bat is *Fledermaus* in German, 'flittermouse' in English dialect, and in various other languages 'leather mouse', 'winged mouse', 'blind mouse' and 'bald mouse'. Bats are actually more closely related to primates than to mice; perhaps that philosophical pipistrelle returns the insult by thinking of humans as bald bears. Speaking of *Pipistrellus*, there are three native species in Britain: common, soprano and Nathusius's. We're also home to two *Nyctalus*, two *Plecotus*, and a couple of stray ones: *Barbastella* and *Eptesicus*.

Anyone keeping tally will have noticed that we still need to account for two native bat species. Both belong to the genus *Rhinolophus* and are better known as horseshoe bats. 'Horseshoe' is quite an ingenious name, based on their facial features, but someone must have run out of inspiration at that point because our British species are boringly known as the greater horseshoe and the lesser horseshoe. A typical lesser horseshoe might weigh seven grams, and a greater horseshoe four times more (that's roughly an ounce in old money). If you're in any doubt about what a horseshoe bat looks like, think of those cartoon bats that dangle upside down from a ceiling with their wings folded round them like a cape.

The two counties where we've lived since getting married – Wiltshire and Devon – are the best for bats anywhere in Britain.

We'd love to be able to claim that our choice of residence wasn't bat-related. Much of Fiona's day job is as an academic researcher specialising in bats. She also gives advice to local planning authorities and government agencies at a European level through the United Nations Eurobats Scientific Advisory Committee. As a side-hustle, we search wind farms for bat casualties with the help of our chocolate Labrador, Charlie Brown. Then there's the behind-the-scenes voluntary stuff: rescuing injured bats and test-flying them round our living room; taking orphaned bats on holiday with us and waking at 3 a.m. to feed them puppy milk; walking cliff paths and country lanes in the middle of the night with a huge aerial trying to find radio-tagged bats; spending countless all-nighters catching bats for monitoring projects; speaking at public inquiries about the damage of new developments to local bat populations; and, most of all, talking with people who are afraid or excited or nonplussed to discover that they have bats in the attic. You get the gist: we love bats.

We have stories to tell about every one of our native bat species, but we've singled out greater horseshoes (*Rhinolophus ferrumequinum*) because they're a wonderful anomaly. All the mammal species that we've discussed so far – beavers, boar, pine martens, water voles – have been flourishing across mainland Europe and struggling here. It's time for us Brits to lift the gloom and give ourselves just a little bit of credit. On the IUCN's European Red List, greater horseshoes are classified as Near Threatened, but that doesn't tell the whole story: they've become extinct in Belgium, the Netherlands and Malta, and their numbers are falling fast in Germany and Austria. They're reported to be doing best in those countries that have limited monitoring programmes, which probably means that they're actually doing badly everywhere – except here.

After precipitous declines over many decades, greater horseshoe populations have stabilised in Britain and are now gradually increasing. They're absent from Scotland (which is beyond their natural range) and Near Threatened in Wales, but in England the species is listed under Least Concern. If ever you needed proof that statistics can be more misleading than lies and damned lies, merely observe that the estimate of 13,000 greater horseshoes in the 2018 population review marks a rise of more than 300% over several decades. While we're happy to take victories wherever we can find them, this recovery is relative to a critically low baseline. Let's choose a different baseline to illustrate the point: according to one estimate, Britain in 1900 may have been home to 330,000 greater horseshoes, which means that now we're left with just four percent of that number. The news isn't actually good; it's just a lot less awful than it was thirty years ago.

It's hard to find reliable evidence of population trends before the twentieth century. The archaeological record is sparse because bats don't fossilise well, and, although they're eaten elsewhere in the world, there isn't much meat on the small species found in Britain, so you won't find bones in middens. The added problem is that everyone took Thomas Nagel's approach: a bat was a bat, and people lacked the will or expertise to discriminate between species. We know that there were large colonies of greater horseshoes in the south-east of England not long before the First World War, but their range gradually contracted to the western side of a line running roughly from Gloucester down to Poole. In Wales, it was touch-and-go for a long time, but the species just about survived at the south-west tip of Pembrokeshire. They hung on wherever hilliness and wetness made arable farming impractical, and where hedges were therefore maintained to control livestock. Greater horseshoes love hedges, particularly the tall overgrown ones that haven't had the life bashed out of them with annual flailing.[1] A thriving hedge is like a tiny remnant wood, providing rich resources for insects (and the bats that eat them) even where there's little prospect of finding anything in the field itself. Now at last the greater horseshoes are expanding east and north in both countries as they start to recolonise old ground. Amidst much rejoicing, a maternity roost has been discovered in West Sussex, and during the last few years individuals have been detected as far afield as Kent.

We can never have enough bat adventures, so in this chapter we're going to describe two very different journeys. The first leads us deep into one of the most mysterious wildlife sites in Britain on a quest to survey hibernating bats. The catch is that you can't follow us there unless you join Wiltshire Bat Group. So we're also going to visit the largest greater horseshoe roost in Cornwall. It was nearby that Thomas Hardy met his first wife, Emma, who later described the setting: 'A beautiful sea-coast, and the wild Atlantic ocean rolling in, with its magnificent waves and spray, its white gulls and black choughs and grey puffins, its cliffs and rocks and gorgeous sun settings.' Those grey puffins are an odd detail: puffins are black and white with orange beaks and orange feet. More to the point – and lovely though it all sounds – what about the bats?

The greater horseshoe is a chunky beast by British bat standards, with a wingspan fractionally longer than a blackbird's. (Blackbirds make up for it by being more than three times heavier.) Catch one fattening up for hibernation and it'll feel squidgy like a well-fed hamster. We think greater horseshoes – with their dense golden-brown fur and pantomime-dame lips – are absolute stunners. Some of our fellow bat-workers dismiss horseshoes as pug-ugly. We deplore their lack of taste.

It's the nose-leaf that divides opinion. A flap of skin forms a horseshoe shape beneath the nostrils, and there's also a small protuberance above the nostrils and a larger arrow-shaped flap, known as the lancet, pointing upwards in front of the brow. Although the dimensions of these features vary between different horseshoe species, their purpose is always to act as sound-directing baffles. Bats aren't blind, and their nocturnal vision is far better than ours, but they've also developed the special skill of echolocating to find food and navigate at close quarters. This involves producing a pulsed stream of sound and listening to the echoes. It isn't unique to bats; dolphins and whales do it, as do a few birds and shrews, and some blind people have learned to use tongue-clicking as a navigational aid. The information bats glean about objects includes their proximity, whether and how fast they're moving, and their texture.

We can listen to echolocation calls using ultrasonic bat detectors. Back in the olden days when we went on our first ever bat walk, we were given a hand-held gizmo that turned bat ultrasound into audible noises. By tuning the dial, we could get some idea of which species were flying around us. Horseshoes were particularly helpful, calling at an almost constant frequency (82 kilohertz for greaters, 112 for lessers) with a bubbly burble that could never be mistaken for the clicks of other species. It was like a car radio: we'd get a perfect reception on the right frequency, and a lot of noise otherwise. Fast-forward a few decades and we now have detectors that enthusiasts can plug into a mobile phone app, sturdy professional-level kit that records unattended for weeks at a time, and fancy computer programmes that take advantage of machine learning and other wizardry to help identify species.

Bat echolocation uses ultrasonic frequencies too high for us to hear. That's just as well because otherwise we'd be deafened: the volume is ten times the legal limit for nightclubs. Bats move apart the tiny bones of the inner ear as they call; if they didn't, the vibrations would permanently damage them. Then, with perfect timing, they move them back together again to hear the returning echo. Horseshoe bats are different from other British species because, although they still do the bone-shifting trick, they keep their mouths firmly shut in flight, and rather than shouting they snort through their noses. They acquire the skill as tiny infants. A newborn horseshoe squawks through its mouth like every other mammal, but by seven days old it's already a proficient nose-snorter. One problem with this particular nuance is that the volume tends to be much lower, and the echo proportionally fainter. The only other nose-snorting British bats are the brown long-eared and grey long-eared. As their names suggest, they have extraordinarily long ears, and these ensure that they don't have to make such a racket. Not only can they pick up the whispers of faint echoes, but they hear the crunch made by a caterpillar on a leaf or the flutter of a moth's wings. A horseshoe bat can also eavesdrop on its prey. Sly scientists have proven the point by fooling it into attacking loudspeakers that are playing the courtship calls of wax moths.

The horseshoe has other tricks up its, er, wing membrane. It flicks the lancet of the nose-leaf and deepens the folds on its face at ten times per second, the same speed as it pulses out sounds[2] – and three times faster than humans can blink. It also alters the shape and orientation of each ear independently, using the twenty muscles that join the ears to its head. (Like most other mammals, we have only four.) These gurning skills focus the bat's ultrasound so precisely that it flies with absolute precision in the confines of a dark cave, and locks onto the circuitous flight path of a moth trying to evade capture. It can even gauge whether to bother leaving its perch to catch a passing bug, having assessed in a split second its size and the availability of better alternatives.

When a greater horseshoe returns to its roost, it chatters away to its friends and neighbours. Judging by their pivoting bodies, wiggling ears and twitching noses, this isn't just idle gossip. We can't yet interpret bat language, but we know that there are established rules of syntax, and at least seventeen different syllables form phrases that are grouped together to make distinct songs.[3] We can sometimes pick up clues from the context; for example, male greater horseshoes have been recorded producing a special trill at cave mouths during the breeding season, probably in a bid to entice passing females to visit. As soon as he succeeds in attracting attention, the amorous male makes short and intense calls to let his intended know that he's hoping to mate; females can reject his advances with bursts of noise that he ignores at his peril – the message is then delivered to him at close quarters courtesy of her pointy teeth. The male's calls seem to be unique to each individual, and they may help females recognise their preferred mate. If he's given permission to proceed, he switches to calls that are four times longer as he inserts his penis.[4] Feel free to provide your own translation there.

What's it like to be a bat? Stand with your arms by your sides. Open your hands, keeping palms forward, and bend your arms upwards and outwards from your elbows until your fingers are outstretched roughly level with your shoulders. Move your elbows slightly out from your body, making a W-shape with your arms. Hold that 'Hands up!' posture. Note that your fingers have grown as long as your forearm. The leading edge of your wing membrane runs from your shoulder to the tip of your index finger; your thumb pokes out of the membrane, the only digit that's not enclosed. The membrane then continues its journey around the tips of the other splayed fingers – the third digit being the outermost point – and comes down to rejoin the body at the ankle.

You're a greater horseshoe, so you're not finished yet. Spread your legs apart. You hadn't noticed until now that you've got a tail hanging down between them. The membrane comes out the other side of your ankle, attaches just above the tip of the tail bone, and joins back up on the inner ankle of the other leg. It's a beautiful design that makes bats faster and more agile than birds. In slow motion they sometimes look like they're rowing through air. To reduce resistance, they can fold their wings during the upstroke, and open them wide on the downstroke. The tail membrane flaps up and down in support, and the knees bend upwards in the opposite direction to ours so as to maximise lift. The trailing edge of the wing doesn't spill air in the way that feathers do; in a remarkable piece of evolutionary jiggery-pokery, hundreds of tiny sensory hairs on its surface mean that changes in air pressure are translated into precise adjustments for aerodynamic performance.

While you're enjoying your impromptu yoga class, you might like to try cloaking yourself in your own wings. Bring your elbows together behind your back, and reach up with your forearms so that your wrists are in line with your ears. Your super-long fingers can now stretch the rest of the wing round the front of your body, where it meets the other wing in the middle. Oh, we forgot to mention that you need to do all of this while dangling upside down from the ceiling. The tail membrane is above you; it stops any drips running down your back by bending forward and turning itself into an umbrella.

When they're resting, most bats like to be squeezed into tight spaces – in tree crevices or rock holes, underneath shutters and cladding, between roof beams, even behind peeling wallpaper in derelict houses. A site inspection to find roosting bats generally involves peering into every cranny, sometimes with the help of mirrors and endoscopes. Horseshoe bats are easier to spot because they often prefer to dangle free. To arrive in that position, somehow they need to land feet first. Horseshoes are extraordinarily aero*bat*ic – sorry! – but they can't reverse while parking, so instead they fly up towards their destination and then turn a last-moment sideways somersault by shutting one wing and extending the other. This is all done in the dark with amazing precision, and their target can be tiny. We've seen horseshoes with their toenails wedged into hairline crevices or clinging to the most filigreed of tree roots that have penetrated through cave ceilings.

It's tricky to judge the size of a bat when it's hanging ten metres above your head; there are plenty of *Father Ted* conversations about what's small and what's far away. People often say that lesser horseshoes at rest look like dangling plums, whereas greaters resemble pears. We think that whether the head is covered or not is a more reliable (though not foolproof) guide: greaters peep out of a small gap between their wings, while lessers are completely sealed up. As for hanging from the ceiling, both species can do it effortlessly. The toes have bumpy tendons that lock like ratchets, holding the digits in a curved position; this is known as a tendon locking system or – not to be confused with the *Times Literary Supplement* – TLS. It takes energy to unhitch the ratchets and let the toes extend. A long-dead and mummified greater horseshoe dangles from a rock in one of our underground survey sites, where it greets us every year. It shows no sign of budging.

Contrary to popular belief, bats can take off from the ground if they have to, but they'd rather not bother. Their preferred method for getting airborne is simply to let go from an elevated perch and start flapping. It takes a bit of falling before effective flight begins, so never put bat boxes in a place with tree branches just beneath the exit hole. Not that bat boxes – at least the conventional sort – are any good for horseshoes. Just as a swallow won't obligingly perch and hop into a bird nest box, a horseshoe bat swoops straight into its roost. Most other bats would land and then crawl, but a horseshoe's quadriceps are short and weak, which prevents it from extending its knees fully. So it crawls reluctantly, awkwardly hoicking itself along by its thumbs.

The area around our old house in Corsham, stretching west towards Bath and south towards Bradford-on-Avon, is riddled with abandoned stone quarries whose names form a quintessentially English catalogue aria: Pictor's, Hollybush, Hazelwood, Monkton Farleigh, Jones's, Box, Henry's Hole, Murhill, Spring, Westwood, Brewer's Yard. 'Quarry' in this part of the world refers not to open-cast stone workings, but to underground sites that range from just a few metres in size to something mind-bogglingly huge. Box Mine, for example, comprises more than eighty kilometres of surveyed passages. From Roman times until the 1970s, honeyed oolitic limestone was hauled out of the ground as building material for Bath and surrounding villages. This work enlarged existing natural features and created roosting opportunities for horseshoe bats in the process. Mines under Wiltshire and Bath are now home to more than ten percent of the national population of hibernating greater horseshoes.

Tim has enough sense (and plenty enough claustrophobia) to keep his cave visits to a minimum, but Fiona reckons she's spent the equivalent of seven months underground. Fetchingly attired in boiler suit, hard hat, boots and knee pads, and armed with a multitude of torches and the all-important stash of Werther's Originals, she meets up for several weekends every year with a small group of like-minded lunatics from Wiltshire Bat Group and disappears down a hole to find the hibernating bats. They've been surveying for more than two decades, and their results give the most detailed information on the social structure and population trends of wintering greater horseshoes of anywhere in Britain.[5] Several of the sites provide a full-body workout, involving boulder-surfing, shinning up rifts and posting yourself through tight holes. 'Do the worm, do the worm,' was the advice once cheerfully called down to our son who, at six feet three inches, was struggling to extricate himself from a narrow slot. Then again, there's something thrilling, almost illicit, about descending into a secret world. Streams of calcite sparkle in your torchlight, and perfectly smooth white cave pearls form in puddles beneath drips. Time flows at a different rate underground. You expect to emerge blinking into the daylight, only to find that it's already dusk.

Soft Bath stone may be nothing like as dangerous as the hard stone mines of Cornwall but, if a slab falls when you're standing below, the result will be – in the words of one of Fiona's Exploration Medicine courses – 'Big Sick'. A caving helmet saves you from a headache when you misjudge the height of a passage; it won't save you from having your head stoved in by a falling boulder. Some quarries have helpful advice etched onto the wall using the black from old carbide lamps. 'No pogoing on the ceiling' appears alongside quips that help attach a date: 'Dave Allen for Pope' or 'Dyslexics against the Pill Tox'. One low slab that's peeling off the roof in Monkton Farleigh is labelled with a warning, but it hasn't moved in thirty years and is now such a fixture that it's marked on the underground map.

'Little Sick' incidents cause difficulties of their own. As landowners aren't keen on legal complications, about twenty years ago they installed grilles at entrances to many of the sites. If you had a good reason for going, you were able to borrow the key for Box Mine from a local pub. That worked well until someone tripped over a rock, broke her ankle and tried to sue the landowner and the pub for giving her what she'd wanted. The landowner's response was to seal entrances and halt access for ongoing bat surveys. Even that wasn't enough to deter the most determined visitors, who can usually find a way in if they look hard enough. Whether they can find a way out is another matter. Several recent rescues have been triggered by people arriving at old exits to find them blocked.

It's startling what you can discover in these underground time-capsules. Names are inscribed on the walls, next to tallies of the stone that's been removed and the payments owing. Drawings mock and caricature the bosses. Patches on the ceiling are blackened by soot from the candles that children once ferried to the quarrymen. There are discarded bottles of ginger beer, cigarette boxes, saws and wobbly wooden haulage cranes; in the larger systems, you can see the furrows made by countless horse-drawn wagons. At Monkton Farleigh, you pass by 'Bugger Hall' and 'King Pillar' through 'Bat Woman's Hole' (named after Fiona – no joke) into 'Gravestone Falls', and find a quarry left very much as it would have looked in the mid-1800s if the workers had just that second downed tools and gone to eat their lunch. Blocks hang half-cut from the walls, and ashlars are piled up ready for loading. This part of the mine was suddenly abandoned after a massive earthquake brought down a large section of the roof, and it hasn't been worked since.

Here's an inventory of other things that we've come across underground: Victorian prams; a rocking horse; the seats from a prototype Mini Cooper; thousands of glass and earthenware bottles; a porcelain-faced doll staring eerily out of a crevice; a metal frog that croaks as you walk by; a Christmas tree complete with baubles, at least two miles from the nearest exit; a shopping trolley (at the end of a crawl about 100 metres long); goalposts from a five-a-side football pitch; golf flags; a century-old lewd drawing of a 'Bathford Flapper'; a large and rather accomplished wall-painting of Satan, wearing what appears to be a white nappy; and an upside-down cross from prank devil-worshipping – at least we *assume* it's a prank – in a space cheerfully marked on maps as 'Death Chamber'.

During the Second World War, many of the mines were commandeered for producing and stocking munitions, building aircraft, and even storing the Crown Jewels and the pictures from the National Gallery. To explain away the sudden influx of thousands of men into rural Wiltshire, the government announced that a major food distribution centre had been established. The mines continued to be used and extended long after the war: the Royal Enfield Works manufactured parts for motorbikes, and the Burlington bunker – a 35-acre government site that runs adjacent to the stone quarries – was dug out in the 1950s to house 4,000 personnel during a national emergency. In the event of nuclear war, the Down-from-Londoners would have poured into their troglodytic new homes, leaving the locals to get cremated above ground. There are cafés, laundries, kitchens, offices, even a television studio for broadcasting morale-boosting messages to the obliterated nation.

The pandemic seems to have increased the footfall through caves; in fact, someone has spray-painted the words 'Covid Lock-In' on several of the Bath sites. We've met families with small helmetless kids on an adventure, urbexers (urban explorers) attracted by the challenge of getting into any site that's locked, and partygoers gathering for events from informal get-togethers to full-scale raves. A great quantity of pot gets smoked underground. We've bumped into a rock band squeezing their drum kit, guitars and amplifiers into Box Mine – they said it would be an awesome recording venue. We haven't yet met the crew making adult films down there – no kink too niche – although Fiona did once deliver a large quantity of heavy-duty chains to a bemused officer at Trowbridge police station. Only a tiny fraction of the people have permission to be there, and the others wander about aimlessly, often with nothing more to guide them than the light from their mobile phones. Lost souls set fire to clothing or pieces of plastic rubbish in an attempt to find their way out. We also come across evidence left behind by groups whose knowledge of caving equipment seems to have been wholly derived from Enid Blyton's *Famous Five* books: bits of string, glowsticks, paint and candles. If you get stuck in a cave system, the best policy is to sit tight and wait – because of course you've been sensible enough to tell someone above ground exactly where you're going and for how long.

In 2016, four teenagers were retrieved from Box Mine by three fire-engine and two rope-access teams. They'd wandered in looking for monsters as part of a Pokémon quest and, lost for hours, they were fortunate to find a phone signal before sunset in the area known as the Cathedral. This is where local businessman William Jones Brewer quarried stone between 1830 and 1850. Brewer had no access rights through other parts of the mine; lacking a horizontal route out, his men hauled stone up to the surface in baskets. As they worked through three levels, they created a huge bell-shaped cavern about forty metres high. The ceiling, which comprises stone and Bradford clay, is reckoned to be only five metres thick, and a row of quarrymen's cottages is balanced above it. The two-metre gap for the haulage baskets remains, letting in the light for which the Cathedral is famous. Happily for those Pokémon players, it also lets in a mobile phone signal.

On a damp November day with energy prices soaring, you'd probably find hibernation an attractive option. Unlike us, bats possess the necessary physiological expertise for the seasonal opting-out, and at a steady 8°C, these limestone caves are the perfect spot for a long rest. (It feels noticeably cooler close to the entrance, which suits species like whiskered bats; they get covered in condensation and end up looking like Christmas tree decorations.) Within an hour of arriving, a greater horseshoe's heart rate plummets from more than 200 beats per minute to just two or three; it becomes cold to the touch and barely breathes.

Endothermy, or warm-bloodedness, is a costly business even for humans. If you get too hot, your enzymes frazzle – and you die; if you get too cold, your enzymatic reactions become sluggish – and you die. The Goldilocks zone takes effort to maintain. You spend about ten percent of your daily energy intake on regulating your body temperature. Another 60–70% of your energy intake goes on similarly non-negotiable tasks like breathing, keeping your heart pumping and letting your metabolism tick over. That leaves only about 20–30% for the luxury of doing anything else. You're currently burning a lot of calories just reading this book; we hope it's worth the energy.

If you're a bat, the situation is even more precarious. Whereas a stranded whale dies out of water because it can't dissipate body heat fast enough to stop itself from cooking, bats have the opposite problem: an awful lot of surface area relative to their volume. Look at a flying bat through a thermal imaging camera and you'll see a fiery red ball, its body radiating masses of heat. As temperatures fall, the costs of staying warm rise. There aren't enough insects around to eat in the winter, so the calories gleaned from hunting simply aren't worth the effort. At a critical point determined by a combination of the weather, food availability and body fat, hibernation mode kicks in. With everything slowed almost to a stop, a bat's stored fat reserves will enable it to survive for weeks, if not months, without food.

We're inclined to think of hibernation as deep sleep, but recent research suggests that the brain activity of a hibernating animal is too suppressed even for sleeping to occur. The idea that animals settle down in late October, setting their alarm clocks to wake up in April, is also wrong: greater horseshoes flit in and out of hibernation. When the weather is good, they emerge to feed. Even in cold temperatures there'll be a few insects on the wing, such as the parasitic wasps that swarm in deciduous woodland and provide important midwinter top-up snacks for greater horseshoes. Dung flies are key species in early winter and early spring, with the bats graduating to crane-flies in April. In the Forest of Dean the bats do something a bit different from other colonies, and feast on the particularly rich populations of dor beetles. These large, dark-blue scavengers clear up woodland floors by dragging animal dung down into underground tunnels. They're beautiful, and evidently very tasty. Backalong, when livestock was routinely turned out into woodland each autumn and winter, these beetles would have been abundant across Britain. Now livestock tends to overwinter indoors, removing dung and its associated bat food from much of our landscape.

Summer brings its own dung-related problems. Juvenile greater horseshoes rely on a diet of dung beetles while their mothers are munching on moths, but the widespread application of toxic veterinary products to rid cattle of worms or parasites kills off just about anything that might live on or in a cowpat. Although a great deal of fuss has been made of the arable revolution and its reliance on synthetic fertilisers and pesticides, the revolution in livestock farming has been just as damaging. Over the ten years to 2018, average herd sizes in Britain rose by thirty-nine percent, necessitating a similar increase in bright green silage fields without a flower or insect in sight. Most pasture is now either grazed flat by sheep and cattle, or cut for silage every few weeks. Farmers cut their grass by April, or early May at the latest, as digestibility falls by about 0.5% a day; some of our local farms start cutting in March if the weather is good. The process is then repeated at four- or five-week intervals, harvesting grass while it's most nutritious. Small wonder that not just insects but also field voles and other small mammals are disappearing fast from the countryside.[6]

But we digress, because before bats can enjoy the summer, they have to survive hibernation. They're primed to wake up when their fat and water reserves are dwindling, but their sensory hairs, which are more sensitive than a human fingertip, are also capable of detecting tiny changes in air flow and humidity. When the weather warms up outside, air incasts into the mines, and it flows in the opposite direction in colder weather. Even humans can feel the difference: we sometimes find our way back to a cave entrance by following the breezes underground. Smaller male bats and juveniles conserve heat by roosting in groups of up to several hundred near cave entrances; the breeze helps them detect temperatures that are likely to support foraging, so they can nip out and hunt when it's worth their while. As winter progresses and weather conditions change, bats move locations within and between roosts. Greater horseshoes are particularly fidgety compared with other British species, and they're on the move once or twice a week for much of the winter,[7] foraging or seeking optimal places for hibernation according to the prevailing weather.

That doesn't mean that disturbance by humans isn't a serious problem for hibernating bats. The heat rising from a human body can rouse horseshoes in minutes. When food isn't available, they've burned a lot of calories for no good reason, and they die if it happens too often. (White-nose syndrome, which has wiped out more than ninety percent of three previously abundant bat species in North America over the past decade, is caused by a cold-loving fungus that disrupts hibernation by irritating the bats' skin, leading to dehydration and starvation.) Bat researchers are careful to limit exposure, and responsible cavers choose routes that avoid bats. The same can't be said of casual visitors, some of whom like to post brightly lit videos of hibernating bats on social media.

Lights and bat roosts are a bad combination all year round, not just during hibernation. Natural England funded a project that tested lighting as a way of persuading Natterer's bats to relocate from one part of a church to another; they had to call it off when the bats refused to come out at all. Even a twilight flier like the common pipistrelle will delay exiting the roost if it's lit up. Long-term research shows that brown long-eared bats stop breeding in churches, and often abandon their roosts altogether, after the buildings are illuminated. The explanation for all this aversion is probably that light makes bats vulnerable to predators. We've seen bats getting picked off by kestrels, sparrowhawks, and even herring gulls, as they emerge from roosts at dusk. On one particular African island where there are fewer avian predators, horseshoe bats show far greater willingness to risk regular daytime flights.[8]

Innocuous though it may seem, street lighting is by far the most important constraint on greater horseshoes' movement through the landscape.[9] For fifty million years, the full moon has been the brightest light encountered by bats, and, thanks to nocturnal vision and echolocation, they've adapted perfectly to a life of darkness. Now lighting has become one of the most pervasive of environmental pollutants. It contaminates roads, rivers and seafronts, spills out of shops and houses, illuminates sports pitches and warehouses, and picks out trees, churches and other buildings for architectural effect. The irony of all this is that humans have rather good night vision until we're dazzled and have to stumble around half-blind for the next twenty minutes. The intensity of outdoor light is increasing fast, with almost no legislation to govern security lights and other fixtures that anyone can buy on the high street. Unless you're materially affecting the character of a listed building, using a tall structure (such as sports-pitch floodlights), or demonstrably causing a nuisance to your neighbours, there's very little that can be done to stop you.

This has serious consequences because bats are naturally peripatetic. In a single year, a female greater horseshoe will visit maternity sites, hibernation sites, mating sites, some transitional roosts (before and after the breeding season), and night roosts where she can socialise with other bats and rest between foraging bouts. Major maternity and hibernation sites are generally well protected by law, but lighting is, so to speak, a grey area. Even trickier is the challenge of safeguarding the numerous small roosts that are also critical to colony survival. For a start, many aren't formally recorded and, even when they've been identified, local authority ecologists have a hard time persuading planning officers that their requests to refuse planning or to modify designs to protect bats are – in the jargon – 'proportionate'. Higher up the chain, the statutory nature conservation bodies are constrained by government policies that prioritise economic growth at the expense of the environment. Yet these small roosts can be pivotal; for example, night roosts let bats put their feet up (quite literally), and give them the chance to forage in places that would be too far away if they had to reach them in one hop from their day roost.

Mating roosts are the most easily overlooked, but they're fundamental to the health of the population. Our greater horseshoes are much more inbred than populations in mainland Europe and, because they're cut off from their continental cousins, they have no prospect of indulging in the kind of exogamy that would fix the issue.[10] Fiona once found a mating site under the terrace of a Grade I listed building owned by a Grade II listed celebrity. There was a male horseshoe in the west wing and another in the east, each living the high life. But it's not all glamour. She's also found a mating site in an old tunnel previously used for growing rhubarb, and a couple in small caves. Here and abroad, very few are known, not least because it takes some work to identify a mating site. There are no louche furnishings, or French movies playing on a loop – just a few bats hanging out and doing nothing much. Study the animals over time and you realise that although the male is a constant fixture, there's an ever-changing cast of visiting females who take themselves away on little family sorties together – mothers, cousins, sisters and aunts – and check out the local talent. If access to a variety of mating sites is lost, whether through lighting or other kinds of development, groups of females end up mating with the same male (sometimes across multiple years) and genetic diversity gets eroded even further. At one roost, just three males had fathered at least a fifth of all the offspring born over a decade, with potentially disastrous consequences for the colony's long-term survival prospects.[11]

A developer flinging up 100 houses probably won't be destroying a maternity site in the process, because strict regulation is in place to protect them. All the same, that housing estate and its lighting can disrupt flight paths so that bats are cut off from feeding habitats or roosts. Existing landscape features get removed, with the promise that something better will be created in their place. That often fails to materialise and, by the time anyone makes a fuss, the developers have finished building, sold the site and scarpered with their swag. Building inspectors, trained to judge whether the correct colour of brick has been used, often lack the expertise to say anything about the appropriateness of the hedgerow. Meanwhile, ecologists working for cash-strapped local authorities are pretty much powerless to intervene. Court cases are costly and time-consuming, and major developers employ lawyers whose daily rates exceed the monthly pay of the opposition.

Given the difficulty of rectifying failures, it's better to exclude key areas from local development plans. To find out which sites are connected at different times of year, we give the bats permanent identity markers. A small aluminium alloy band like a clip-on bangle is slotted over each bat's forearm in a way that ensures it can slide around a little but won't drop off. We've also started using coloured rings that are identifiable on bats even when they're too high up in a roost to read the number. Within a year of trying this new technique, Fiona found that bats hibernating in a small cave not far from Plymouth were also frequenting one of the major designated maternity sites and a smaller maternity colony some miles away. When we map these networks, we can begin to protect the landscapes that link them.

The downside of ringing studies is that they only provide snapshots. GPS collars of the sort now regularly attached to lions and birds of prey would offer much more detail, but they're heavier than the bats that we want to track. The next best thing is a tiny VHS transmitter that's attached to the bat's back with the same glue that's used for false eyelashes. It stays in place for a couple of weeks, all the while transmitting a beep that can be picked up by receivers. With a small army of volunteers, we've scoured the countryside in the early hours, trying to locate radio-tagged bats. Several times the police have stopped us to ask what we're doing, and they immediately believe us because no one would ever make up a story like that. Radiotracking has rewards as well as frustrations: there's a lot to be said for watching the sun rising slowly above the Jurassic Coast, and even balancing on a five-barred gate on a pitch-black night in a desperate attempt to pick up a signal has its comic upsides. But you can't be everywhere at once, the roads are twisty, and bats fly fast; all too often, we never see them again. In winter, you can sit all night outside a cave without a single bat emerging to feed. So we're now installing a new system across Devon that should solve the problem. Instead of wild bat chases, we monitor flight paths from the comfort of our living room, thanks to a network of radio-receivers that automatically log bats as they fly by.

A batch of batty facts to combat insomnia:

In Leviticus 11, bats are identified as birds alongside hoopoes, owls, raptors, storks and various others. They're pleased about the error, because this happens to be a list of animals that are 'unclean' and shouldn't be eaten. Even so, in Sulawesi, flying foxes (large fruit bats) are eaten in huge numbers at Christian holidays such as Easter and Christmas.

The seventeenth-century naturalist John Ray decided that bats couldn't be birds because they give birth to live young. He grouped them with 'the four-feet animals'. In 1758, Carl Linnaeus was the first to classify bats as mammals, although he mistakenly concluded that they must be primates. At least he didn't call them mice. One factor in Linnaeus's decision was the high position of the nipples, which had already been noted in some Christian teachings as a sign that bats were demonic. In South-West Asia, the same feature has traditionally counted in their favour, on the basis that female bats have breasts and are mothers like human females.

The greater horseshoe is one of the few species longer-lived for body size than humans. They can reach the grand old age of thirty or more. Fiona's oldest greater horseshoe bat record was Boris, whose naming preceded the rise and fall of our Glorious ex-Leader. She was part of a group that found him down a mine in January 2000. He was already twenty-seven or twenty-eight, making him a mere whippersnapper compared to another of our native species, Brandt's bat, which can live into its early forties.

The Mexican free-tailed bat is the world's fastest mammal; it flies at speeds of up to 160 kilometres per hour. Greater horseshoes are visibly slower in flight than many bat species, with a top speed in the region of thirty kilometres per hour. They compensate by being extraordinarily manoeuvrable as well as lazy: they often wait on a perch for an insect and drop down onto it.

Bats are one of the few mammals where the females are usually larger than the males. Bigger males have more offspring than smaller ones, whereas even puny females will breed in most years. Unlike most other social animals, bats are an egalitarian bunch with no hierarchies.

Male bats have genitals that, obviously allowing for size, at first glance look similar to human ones. Less obvious is that most of them, including all British species, are proud possessors of a baculum – a penis bone – which is unconnected to any other part of the skeleton. Humans mislaid their penis bone somewhere over evolutionary time. (God apparently made Eve out of Adam's *rib*, so that can't be to blame.) We're the only primate to go without a baculum, and biologists have had endless fun speculating about why. Maybe sex doesn't last long enough to make one worthwhile, or maybe (definitely!) you don't want a penis that can be broken by rivals. These seem like plausible if rather deflating explanations.

Many bats are also proud owners of penis spines. These protuberances are made from keratin and vary in shape and size, ranging from a little goose-bump to something more eye-watering: were the North American hoary bat the size of a human (and, yes, our human ancestors also had prickly penises), its spines would be ten centimetres long. One probable function of penis spines is to stimulate the female, but they can also act like a bottlebrush to clean out a rival's sperm when a male is late to the party. After bats mate, the semen and vaginal secretions harden to form a vaginal plug, which is a bit like a waxy tampon. This plug continues to enlarge over the following few months, and stays in place until just before the baby is born – unless another male comes along and uses his spines to brush it out of the way. It seems that frisky females can eject the plug all on their own if a new male takes their fancy. In case this weren't already more than enough, several bat species are the only animals other than primates known to enjoy oral sex. At least they look like they're enjoying it. Voyeuristic scientists with stopwatches have shown that whether it's cunnilingus or fellatio, oral sex makes bat coitus last longer.

Birth must somehow be made to coincide with peak food supplies, which in Britain means that the egg should be fertilised around the end of March for a three-month gestation. There's little chance that a male can make sperm at that time. He's likely to be hibernating, and, even if not, food's in short supply. So male horseshoes have evolved the capacity to store live sperm for months. Sperm can last in the epididymis well into the winter, and it's still in good shape for mating during January and February; that compares favourably with human sperm, which survives in the epididymis for a couple of weeks at best. February would still be too early for mating, were it not that female horseshoes are among a very select group of mammals capable of storing sperm. The human female reproductive tract is an inhospitable place for sperm, which can last for no more than five days; a female greater horseshoe actively nourishes sperm through the winter. This gives her the option of mating in the autumn and fertilising her egg six months later. Rather than being at the mercy of whichever male chooses to take advantage while she's hibernating and too torpid to resist, a female greater horseshoe can select the suitor that she would prefer to father her one precious baby each year.[12] The likelihood of successfully producing offspring increases year on year until a female horseshoe turns twelve (fourteen in males), after which it declines gradually. Some males are still successfully siring babies at the ripe old age of twenty-four.

If climate change predictions prove accurate, Britain can expect cooler and wetter springs over the coming decades. Pregnant and lactating bats find these conditions particularly difficult:

food supplies are low, and they can't become torpid because foetal development and milk production will be affected. One way or another, they have to stay warm. Temperatures in April and May seem to be critical for pregnant greater horseshoes, and birth can be delayed by several weeks in years with poor springs. This gives the young bats less time to gain the body fat needed to survive hibernation. When cool weather continues into the summer, pups spend their energy staying warm rather than growing, and that can leave them permanently stunted. They need to have reached ninety-five percent of their adult size by six weeks of age, with bones stabilised to withstand the demands of flight, so there's no hope of catching up if they miss out on that crucial developmental phase.

One of the main priorities for maternity colonies is therefore to find locations that are reliably warm. Born furless, for the first fortnight a baby bat is incapable of regulating its own body temperature, so it snuggles under the mother's armpit and keeps clinging on. This arrangement isn't exactly aerodynamic, so horseshoes have false nipples – two flaps of skin near the genitals – to which the pup attaches itself during flight for better streamlining. But trying to catch moths on the wing is difficult enough at the best of times, and a newborn pup is already between a quarter and a third of its mother's weight. (For comparison, a human baby is around 1/22nd of the weight of its mother, and a tiger cub 1/120th.) Sooner or later, the situation is unbearable; the baby gets left behind in the roost and the mother returns several times in the night to feed it.

An isolated infant in a cool roost will quickly become hypothermic, which is why the first of our emergency actions whenever a baby bat is brought to us is to give it a hot-water bottle. Babies left by their mothers in a roost gravitate towards each other, clustering for warmth like penguins on an ice-floe. One or two adults often stay behind, minding the crèche. That's quite a challenge in large roosts where the numbers of babies can run into the hundreds. Babies that fall to the ground have little chance of survival unless their mothers retrieve them, but, luckily, mothers and infants have a unique pattern of alternating calls that only fit together when they've identified each other. Having found her pup, the mother envelops it in her wings and emits a special contact or comfort call.[13]

We know from the Vincent Wildlife Trust – them again! – that the keystone for any bat conservation strategy is the protection of roosts. As greater and lesser horseshoe numbers were collapsing towards extinction in Britain and Ireland during the 1990s, the VWT made the decision to start purchasing buildings with roosts. It manages about forty of them, and the population increases at its sites far outstrip those seen in other locations.[14] Around fifty percent of the British greater horseshoe population is accommodated in its six breeding roosts.

Some of the methods developed by the VWT are now widely adopted. One of the first essential steps for enhancing a roost is to reduce draughts. This doesn't have to be high-tech: in a natural cave roost at Cheddar Gorge, bat-workers achieved the same result by stuffing an old mattress across part of the cavern entrance. It also helps to split roof spaces into sections, install mezzanines and attics into open barns, and fix insulated wooden boxes ('hotboxes') in the apex of roof spaces. These measures focus on keeping the roost warm, but occasionally

– and increasingly – there's the opposite problem: baking summer temperatures can cause bats to die from heat exhaustion. British researchers are starting to provide bats with flexible living arrangements according to weather, offering cool cellars as respite from extreme heat.

Greater horseshoes are happy to accept modifications, but they're more resistant to the idea of moving house entirely. Given the option of relocating to a purpose-built des-res roost, one colony in Dorset refused to abandon their draughty accommodation. Lesser horseshoes seem to be more pioneering, and they'll willingly squat in a building due for renovation if a door is left open or a windowpane is smashed. In 2017, a small colony delighted national media by taking up residence in the belly of a fibreglass triceratops at Combe Martin in North Devon. In our time, we've been called out to find lesser horseshoes hanging from the drapes of a four-poster bed, in the library of a stately home, within the upturned bell of Saint Winifred's Church in Branscombe, and in the 'grotto' (complete with tasteful stalactites imported from China) of an indoor swimming pool owned by a television celebrity.

It was summer and we were visiting Boscastle – one of the loveliest Cornish villages. Things must have improved since John Leland dismissed it almost 500 years ago as a 'very filthy toun and il kept'. Leland even took to moaning about the 'bleke northen se', which does indeed hammer Boscastle's natural harbour with an elemental fury when the winds and tides conspire. In 2004, the village was hit by flash floods after monsoon rains; 150 people had to be winched to safety by Sea King helicopter. You don't live in Boscastle unless you like weather.

When his wife Emma died in 1912, Thomas Hardy came to this stretch of coastline to relive the scenes of their courtship from more than forty years earlier. It's fair to say that the marriage had soured somewhat: among Emma's papers, he found a set of diaries titled 'What I Think of My Husband', and decided after reading them that the best thing all round would be to throw them on the fire. The elegies inspired by his nostalgic visit to Cornwall veer between grief, love, accusation, guilt and bewilderment. The backdrop is a landscape at once permanent and in flux, with its 'primaeval rocks', 'blind gales' and 'salt-edged air'. We were in Boscastle mostly for the bats, but some of the reasons why they're doing so well are closely related to the geology and climatic conditions that Hardy's poetry celebrates. Horseshoes enjoy maritime environments because they're likely to be remote, less intensively managed, and protected from extreme temperatures. The bats roost in the Valency Valley half a mile above Boscastle, and they benefit from the shelter provided by steep tree-lined banks. On a summer night, colder air pools in the valley and the hilltops feel noticeably warmer. The bats follow the food supply, so they move uphill as the night progresses. Their home is Minster Church, where at least 200 female greater horseshoes set up their maternity roost each summer. It's in a secluded spot above the village, and both the building and its surroundings feel properly ancient. Because of the holy well nearby, there's been a church on this site since early in the sixth century, and what we see today is Norman with some unhelpful Victorian repairs.

Churches aren't always sympathetic to bats, although the Movement against Bats in Churches (known by the ungainly acronym MaBiC) seems to have fallen dormant after several years of lobbying for permission to exclude them from sensitive religious sites. Even a bat-lover can appreciate the frustration felt by campaigners who (in a very small number of churches) saw medieval murals damaged by bat wee, or (in many others) didn't care for the bat droppings liberally scattered around fonts and choir stalls. At one point, it looked like theologians would match the philosophers by coming up with their own urgent question for our times: is a bat capable of desecrating an altar?

The truce currently in place reflects the success of mitigation measures to reduce the impact of bats, as well as some excellent PR work by the Bats in Churches project that brought together the Bat Conservation Trust, church authorities, parishioners and government agencies. We needed a solution urgently, because protected bat species often have nowhere else to go, having lost their roosts to tree-felling, barn conversions and other developments. Whereas policy on biodiversity in farming is starting to move in the right direction – public money is given for public goods – this still doesn't apply to bat roosts. A local church may be home to all the bats for miles around, yet there's no earthly reward for its hospitality. The greater horseshoes at Minster Church are fortunate that the authorities have made a virtue of their bat roost, and they manage their land so sensitively that it's been designated a Site of Special Scientific Interest.

We were staring up at the saddleback tower, all five of us, each with our own bat detector. You can never quite predict when the magic will start. So much depends on the temperature, how well the bats ate the previous night, the phase of the moon, and when and whether they can be bothered. We'd guessed correctly that this colony would emerge earlier than most because there was very little light pollution and plenty of shade from surrounding trees. Then it began – a flittering blackness against a black sky. A couple of bats very kindly did some circuits around the tower, and one passed several metres in front of us as we stood on the path below. Most headed straight down the valley, in numbers that were driving our bat detectors loopy. We lost count at somewhere over a hundred. What we couldn't see but only imagine was that many had pups clinging to their fur. And it got us thinking: what's it like to be a greater horseshoe pup, hanging on for dear life? Are they exhilarated? Are they terrified? Or is it just ordinary, no big deal? If only the philosophers could tell us.

SIX

Tiggywinkle Goes Rogue

On the evening of 10 November 2015, Rory Stewart MP, the then Parliamentary Under-Secretary of State for Environment, Food and Rural Affairs, rose to the despatch box at the House of Commons. He spent the next thirteen minutes holding forth on the subject of hedgehogs. When he sat down again to cheers from the dozen or so MPs who'd bothered to attend, Madam Deputy Speaker declared that he'd 'just made one of the best speeches [she'd] ever heard in this House'.

Watching the performance on YouTube today, it's hard not to marvel at that innocent age before Brexit, Covid and war in Ukraine, when politicians were still allowed to be frivolous. And Rory Stewart was nothing if not frivolous, as he produced quirky fact after quirky fact about what he called a 'magical creature'. His fellow Tory MP, Oliver Colvile, had requested the debate in the first place, having been shocked to discover that hedgehog numbers had declined by about a third in just ten years; Colvile proposed that, as a way of forcing this crisis into the public consciousness, we should promote the hedgehog as our national symbol. Ignoring altogether the serious part about the population crash, Stewart wound up his peroration with a polite but firm refusal. How could we adopt as our national symbol an animal that spends six months of the year asleep, and that 'when confronted with danger rolls over into a little ball and puts its spikes up'? Stewart's preference was for retaining the incumbent: 'I refer, of course, to the lion, which is majestic, courageous and proud.'

That the minister responsible had nothing to say about the gradual disappearance of a much-loved species should come as no surprise. For hundreds of years, wild animals have taken up parliamentary time only when they're posing a threat to our safety or our food supplies, or when there's a skirmish over the right to hunt and kill them. Governments of all stripes have lavished wild animals with indifference. Stewart himself noted that hedgehogs had previously attracted the attention of parliament on only one occasion: in 1566, when a tuppence bounty was put on their heads (or, more gruesomely, their severed snouts), partly because of the delusion that they stole milk by sucking on cows' udders. That woeful law resulted in the death of more than two million hedgehogs.

At least Colvile's actions were well-intended, even if their primary effect was to advance Rory Stewart's burgeoning reputation as a national treasure. Unfortunately, the lesson of the entire episode was that it's pointless to hold forth about magical hedgehogs while doing nothing to prevent their vanishing act. All the evidence suggests that hedgehog populations have continued to decline since Stewart's speech. There's been no government plan, no intervention, not even so much as a working party. A 'Species Champion' for hedgehogs has, admittedly, been appointed at Westminster; bragging rights if you can name him. (All will be revealed in due course.) Unless something much more drastic is done, when parliament gets round to discussing hedgehogs 450 years from now – or in fifty years, for that matter – it'll be in the past tense.

Where do you go to see a hedgehog? Many other species that we've headlined have distributions that are patchy and fragmented. As things currently stand, you won't find a wild red squirrel in Kent or a pine marten in Nottinghamshire, no matter how hard you look. Hedgehogs, on the other hand, range across the entirety of mainland Britain. So keen are people to report a hedgehog encounter that we've got more records for them than for any other mammal species. We can therefore be pretty confident that hedgehogs are still present in almost every ten-by-ten-kilometre grid square, much as they were in 1995 when the population was previously assessed. They're all around us, in rural and urban areas. They bimble through the churchyard fifty metres from our front door here in Colyton. Our friend Andy regularly encounters them on his early-morning postal rounds. Social media is full of trail cam footage of hedgehogs feeding in people's gardens. Yet hedgehogs are in serious decline. They're becoming extinct in plain sight.

The 2018 population review suggests that there's been a fall in hedgehog numbers of two percent per year, although this comes with a flashing red light. The population probably stands at about 880,000 individuals, but we can't place that estimate within plausible intervals because for many habitats we lack basic information: there's an outside chance that some woodlands are still stuffed to the gunwales with what is, after all, traditionally a woodland animal. The estimate made back in 1995 was just as shaky, so when we try to work out trends in population size over time, we're comparing one best guess with another. Different kinds of studies argue that things could be much worse than we've assumed. In recent years, there have been five systematic surveys in which volunteers have visited a particular area and reported whether or not they saw a hedgehog. There's variability in the size of the change, but the direction is always downwards, giving an estimated decline of about four percent per annum.[1] Putting all the evidence from the population review and the citizen science studies together, the Red List concludes slightly less pessimistically that the decline each year is running at a little under three percent. There being no reasonable prospect of reversing the trend in the near future, hedgehogs have earned their Red List classification of Vulnerable across Britain, with an additional recommendation that the status should be regularly assessed.

One of the most compelling reasons to assume that hedgehogs are struggling is that fewer are being squashed on our roads. After Tufty the squirrel hung up his blue jacket in the 1990s and retired to a Scottish pine forest, the government-backed road safety campaign 'Think!' appointed in his place a family of animated hedgehogs. Hedgehogs continued to groove their way across our screens to the tune of 'Stayin' Alive' or 'King of the Road' for almost two decades, by which time spotting a flattened hedgehog had become a rare event. Had hedgehogs started paying attention to their own televised warnings? Or was this an example of evolution in action, with hedgehogs that were genetically inclined to run outsurviving and outbreeding their roll-into-a-ball cousins? Perhaps over time this kind of natural selection is possible, but it's unlikely to be the primary explanation for the dwindling numbers of squedgehogs reported in the People's Trust for Endangered Species' annual 'Mammals on Roads' surveys. Traffic on major British roads has increased by about ten percent since 2000, so a stable population of hedgehogs should see a proportionate increase in casualties. The decline suggests that there are fewer hedgehogs available to get themselves squashed, although the toll is still bad enough: according to one estimate, somewhere between 100,000 and 300,000 hedgehogs are killed on our roads every year.[2]

So it's grimly ironic that the first Species Champion for hedgehogs, Chris Grayling, took on the role while still in post as Transport Minister. During his time in office, he was keen on funding new roads – £1bn-worth of them. Weigh that hedgehog-squishing investment against his launch in 2019 of a 'hedgehogs crossing' traffic sign and you'll appreciate the priorities. Tipped off in advance that the transport ministry was looking to promote road signs as a good news story for conservation, Fiona's research team, funded by the Mammal Society and the People's Trust for Endangered Species, had already been researching road casualty hotspots.[3] They showed that risks to hedgehogs could be predicted by habitat: areas with around fifty percent of urban cover and plenty of grassland make up only nine percent of the total road network, but are far and away the most dangerous places. The resulting interactive maps of road casualty hotspots are freely available on the Mammal Society website. Researchers are still finding out whether particular features, such as a grass verge or hedgerows meeting a road at right angles, encourage animals to wander into the paths of cars.

All of this raises a question: even if we know where hedgehog road casualties occur, do traffic signs work? Very few local authorities have applied for permission to use them, whether because they have doubts or they just don't care enough. Fiona once reported a hedgehog road-crossing hotspot to East Devon District Council, only to be told that the matter wouldn't be taken further because Computer Says No: checks of the database hadn't found any human casualties from road traffic accidents involving hedgehogs in that area. For the council to act, a hedgehog needed to bounce across a car bonnet, smash through the windscreen and embed its spines in the driver's eyeballs. Less obtuse councils are still wary of driver information overload: we know that signs warning of danger to motorcyclists are ignored if too common, and no longer seen if kept in place for too long. Some thought has been given to hedgehog bridges and tunnels, but the cost of adding them to existing roads is prohibitive. For new roads, effective wildlife crossings require a small number of well-defined points where animals might take their chances. Hedgehogs don't behave predictably enough. Anyway, erecting fences may work for a major trunk road, but it's hardly going to be appreciated by a frail pensioner having to go the long way round to the shops on a Sunday morning.

Learning from the 'ghost bike' project, whereby every cyclist killed on the road is memorialised by a painted white bicycle chained near to the accident site, the Dorset Mammal Group has started putting out 'ghost hedgehog' signs. We first noticed them one evening on a roundabout at Blandford Forum, the white silhouettes illuminated in our headlights. Officialdom doesn't care for roadside distractions, but if the net result proves to be a reduction in hedgehog casualties, this guerrilla approach should be adopted across Britain. They're visible, attention-grabbing and seasonal – they can be removed in winter when hedgehogs are hibernating so they don't lose impact through familiarity. Despite the government's much-advertised push for evidence-based policy making, we're not aware of the Department of Transport (or the Statutory Nature Conservation Authorities) currently funding research to test whether hedgehog-crossing signs or ghost hedgehogs do anything to reduce fatality rates.

Dare to mention hedgehog declines on social media and, within minutes, your inbox will be brimming over with bloodcurdling declarations of war on another of our native species: badgers. When it comes to saving hedgehogs, killing badgers seems to be a much more popular strategy than giving up driving. Full disclosure: Brock not only competes by eating the same foods as hedgehogs, but will also occasionally take the direct approach and eat the hedgehog itself. Having researched badgers for years, Fiona can guarantee that you don't want to get on the toothy side of them. If you're bitten by a badger, you stay bitten. Their jaw muscles are the strongest of any terrestrial mammal in Britain, and their claws are plenty sharp enough to unzip a hedgehog. In one well-known study, hedgehogs were observed swimming across a river to escape their stripy nemesis;[4] badgers can *kind of* swim, but seem reluctant and rarely bother. Hedgehogs will go to great lengths to avoid the scent of badgers, which is possibly why they're less likely to be found in areas with high badger densities. (It's not simply that they've all been eaten.) Badger numbers are also on the increase, bouncing back from centuries of human persecution now that they're protected by law.

The final clue pointing to badgers as Tiggywinkle-chomping ruffians comes from the randomised badger culling trial, a project set up to investigate whether culling badgers is an effective way of managing bovine tuberculosis. (It isn't, but the government carries on regardless.) This reported that hedgehog densities in amenity grassland such as parks or playing fields increased in areas where badgers were culled (from 0.9 per hectare to 2.4 per hectare), whereas they remained unchanged in control areas (at 0.3 per hectare).[5] So that's fantastic: all you have to do to stop bovine TB *and* save hedgehogs is kill hundreds of thousands of badgers. On closer inspection, these statistics aren't clear-cut. The data came from just twelve fields of amenity grassland (local sports pitches, parks and so on) in cull areas, and twelve in control areas without culling, and these were clustered into four study zones. Three times as many pasture fields as amenity grassland fields were also surveyed, but the numbers of animals seen were so small that the data were discarded altogether. The evidence is therefore taken from locations fairly unlikely to have badgers in the first place, rather than where badgers are abundant.

Does the science provide unequivocal evidence that badger culling would improve the conservation of hedgehogs? No it doesn't, because with such low replication, confounding effects become important. Let's consider the possibility that someone starts to leave food for hedgehogs in a cull area, or that there's a favourable change in habitat management. That would be sufficient to explain an extra couple of hedgehogs even if badger culling had no effect at all. It's not to deny that badgers are important predators. No rehabilitated hedgehog would thank you for releasing it adjacent to a badger sett, but at a population level the impacts of badgers are likely to be small. Badgers and hedgehogs have coexisted for millennia in Britain and across Europe. If badgers were responsible for the decline, they'd have munched hedgehogs into extinction a long time ago.

No writing about hedgehogs can be complete without mentioning that they're the nation's favourite wild mammal. They win poll after poll after poll. What kind of monster is uncheered by a hedgehog? They're small, they're cute, they're adorably weird, they eat creepy-crawlies, they're familiar yet mysterious, and the impression that they're ill-suited to modern life somehow makes them even more endearing. There are reportedly more than 600 hedgehog carers in Britain. The massive practical knowledge developed by people like Les and Sue Stocker, founders of the famous St Tiggywinkles wildlife hospital in Buckinghamshire, is complemented by hedgehogophile authorities such as Hugh Warwick and Nigel Reeve, who know *everything* about the species

– from how many spines hedgehogs have (somewhere between 3,500 and 7,000 apparently, although we haven't checked), to how fast they can run (nine kilometres per hour, which is roughly the same pace as the average parkrunner). One way or another, Britain is a hotbed of hedgehog excellence. The animal's eccentricity seems to speak to something diffident and self-deprecating within our own national character; according to a YouGov poll, six out of ten Britons describe themselves as 'reserved', and you can't get much more reserved than a hedgehog. In the sporting arena, three hedgehogs on your shirt may not convey quite the right message, but under any other circumstances leonine swagger seems faintly repellent compared with the shy and droll ways of the hedgehog.

Whisper it, but there was, actually, one rather curious individual who didn't love hedgehogs: William Shakespeare. Audiences search in vain for his opinions on almost any other issue; he is, as Keats pointed out with some wonky spelling, 'the camelion Poet' who, being 'everything and nothing', takes 'as much delight in conceiving an Iago as an Imogen'. That's all well and good, but Shakespeare's principled impartiality collapses at the very thought of hedgehogs. Imagine that the person who killed your father has just unrepentantly revealed his crime to you. What insult would do justice to your rage and their depravity? In *Richard III*, Anne's response at that very moment is to denounce Richard as a 'hedgehog'. When the fairies of *A Midsummer Night's Dream* sing a song to protect Titania, they address the scariest dangers that threaten her: 'You spotted snakes with double tongue, / Thorny hedgehogs, be not seen'. Caliban in *The Tempest* reports the punishments he must endure after every minor transgression: spirits take the form of 'hedgehogs which / Lie tumbling in my barefoot way and mount / Their pricks at my footfall'. Caliban had already complained of 'urchin-shows' – 'urchin' being an older word for hedgehog that survives in its spiky nautical counterpart, the sea-urchin. So an 'urchin-show' is not, as our bells-and-whistles annotated edition of *The Tempest* blandly asserts, 'a goblin-show', but the malign appearance of spirits in the form of hedgehogs. Caliban's master, Prospero, had threatened him earlier with the promise that 'urchins / Shall, for that vast of night that they may work, / All exercise on thee'. These examples are too frequent, and somehow too self-exposingly strange, to be dismissed as coincidental. We may not know a lot about Shakespeare's character, but we do know that his most intimate fear was to be prickled by a hedgehog. Bear that in mind next time you're awed by the philosophical profundity of *Hamlet*.

Perhaps young Will had become traumatised after falling on a hedgehog while out apple-scrumping one night, or perhaps he was merely influenced by a widespread suspicion of hedgehogs and their motives. The Irish for hedgehog is *gráinneog* – literally, ugly or horrible one, as befits an animal often suspected of being a disguised witch. Hedgehogs were, at the very least, animals of ill omen, and when the witches in *Macbeth* start an incantation around their bubbling cauldron, they invoke the hedgehog and 'brinded cat' as their familiars. How much the 1566 legislation provoked such calumnies, and how much it was itself a product of public opinion, may now be too tricky to unravel. Either way, it's a handy reminder that when Shakespeare was writing, hedgehogs were widely seen as bandits, stealing milk, grain, fruit and eggs under cover of darkness.

Some sense of their elevated place in the hierarchy of vermin during the sixteenth century can be garnered from the relative largesse of that tuppence reward for killing a hedgehog; polecats and stoats were worth a single penny, and the brilliantly named moldewarpe (mole) a mere ha'penny. It must have been easier to catch a hedgehog than those other species, and the farmhand's average daily wage of sixpence made the effort relatively lucrative. There were food riots in the 1590s after a series of disastrous harvests, and many people starved to death. Shakespeare may have been an outlier in the idiosyncratic intensity of his loathing, but with all the misinformation and the old wives' tales counting against them, hedgehogs at that time wouldn't have been winning any polls as the nation's favourite mammal.

In fact, prejudices seem to have persisted for hundreds of years, long after people should have known better. There are very few positive accounts of hedgehogs before the twentieth century. Writing in the 1830s, the naturalist-poet John Clare deplored the ongoing persecution, unhappily concluding that 'no one cares and still the strife goes on'. Even after the Elizabethan law was finally repealed in 1863, attitudes and behaviours seem to have continued as before. It needed something remarkable to transform public opinion. Step forward Beatrix Potter, who in 1905 published *The Tale of Mrs Tiggy-Winkle* about a hedgehog washer-woman with a sniffly nose and twinkly eyes. *Mrs Tiggy-Winkle* is, frankly, too mawkish to be one of her best books, but it's certainly among her most popular. Only three years later the Ministry of Agriculture, possibly reacting to a sudden surge in hedgehog love, issued a notice to farmers reassuring them that hedgehogs didn't, after all, steal milk from cows. That advice came with no prohibition attached, but in more recent times the Wildlife and Countryside Act of 1981 and its subsequent amendments have made it illegal to capture or kill wild hedgehogs, and the Wild Mammals (Protection) Act (1996) forbids cruel treatment. No doubt they're killed, both accidentally and sometimes on purpose, by gamekeepers protecting the eggs of ground-nesting birds like pheasants and grouse, but not in sufficient quantities to explain the crisis. It's cold comfort for hedgehogs that their continuing decline now has very little to do with direct human malice.

One reason to feel gloomy about the prospects for hedgehogs is that there's no single clear cause to address. The campaign to put gaps in fences so that they can move between gardens has gained massive traction in the press and on social media. It may even work: hedgehogs certainly do use the holes on occasions. The unpalatable truth, though, is that catastrophic population decline is not owing to a lack of access between gardens, especially when we consider that hedgehogs seem perfectly happy to trundle along suburban streets and up and down driveways. If we knock a hole in our fence and assume that hedgehogs will be fine, we're complacently condemning them to extinction.

A major issue, partially addressed by those fence holes, is prey availability. The obsession with tidiness, the paving of driveways, the wretched proclivity for plastic grass (with which the ninth circle of the Inferno is probably carpeted), and the human intolerance of overgrown land where beetles and bugs can breed all play their part in the hedgehog's inexorable decline. In the week of the COP26 summit in Glasgow, when the world's attention was focused on climate change and biodiversity loss, our parish council sent in contractors under cover of darkness to clear a brambly field that provided habitat for three protected species: hedgehogs, slow-worms and dormice. Despite having been warned about the law, they didn't bother to carry out any ecological surveys. The rules are barely policed, after all. This is how biodiversity loss happens under our noses – field by field, hedge by hedge. And (usually) the people who carry out this destruction aren't wicked; they may well wring their hands over the clearing of the Amazon rainforest, but they don't view their own actions as part of the same destructive processes.

There are immediate as well as long-term consequences. In and around Colyton, it isn't uncommon to find emaciated hedgehogs with few, if any, fat reserves. Our younger daughter picked up an underweight hedgehog on an early-December walk home from school – it didn't survive, and a post-mortem examination at London Zoo showed that it had absolutely no fat reserves at all. It hadn't managed to eat enough in the autumn to lay down sufficient energy stores for hibernation. This could have been because it was born very late in the season and ran out of time before winter set in; hedgehog rescue centres are overrun with 'late juveniles'. There's also evidence that some of these small animals are born at the normal time of year but have failed to grow properly, presumably through lack of food.

Evolutionarily speaking, hedgehogs are mostly insectivorous. Their diets can vary considerably depending on location, the season, and the age of the animal, but invertebrates are their main food, supplemented with small vertebrates, carrion, and the occasional nut, mushroom or fruit. Other plant material such as grass commonly occurs in hedgehog droppings, but it doesn't appear to be chewed or digested, having probably been eaten by accident. The hedgehog is particularly partial to noctuid moth caterpillars (cutworms). These feed on grass and agricultural crops, spending the day hidden in the soil and coming out to feed at night when they're safe from birds. But they're not safe from hedgehogs: two separate studies have found more than fifty-five individual caterpillars in a single hedgehog dropping.

The only things hedgehogs find tastier than caterpillars are beetles. A wide range of species are taken, but they're often small- to medium-sized predatory ground beetles. Wherever they can find them, hedgehogs will also feast on larger dung-beetles and cockchafers. The size of a humbug and crammed with juicy innards, cockchafers are the Big Mac of the insect world, but probably tastier and certainly better for the environment. Some hedgehogs eat earthworms, but others rarely touch them. They'll also take earwigs, millipedes, harvestmen, sand-hoppers, spiders and crane-fly larvae (the leatherjackets despised by many lawn-manicurists and allotmenteers).

The reputation of hedgehogs as the gardener's friend therefore seems justified, even if their consumption of slugs and snails is rather overstated. Hedgehogs struggle to break into the thick shells of larger snail species, so the ones they can manage tend to be small. Even these are taken only occasionally. Slugs such as the grey field slug are eaten more often, which is hardly surprising given that it's widespread on both grassland and crops. It looks innocuous enough, with blotchy grey patterning and distinctive milky mucous, but the website of the agrichemical giant Bayer tells us that it's 'the most common and significant slug pest in oilseed rape throughout the UK'. (This argument is arsy-versy: the pest here is oilseed rape itself. Its cultivation leaches vast quantities of nitrates into our watercourses, and it's so heavily reliant on pesticides that it sits at the centre of debates around both neonicotinoid insecticides and genetic modification of crops.) Should we, then, be releasing hedgehogs into fields as a way of keeping the slugs down? The available evidence, of which admittedly there isn't much, indicates that slugs make up between one percent and six percent of their energy intake, despite being some of the most abundant prey species available. So hedgehogs don't appear to be especially enthusiastic. They've even been observed wiping slime off the slugs before eating them – a sure sign that they find them a bit gross.

Although hedgehogs are most frequently spotted in short grass, that's partly because they're easier to see there than in more overgrown areas. They can cope perfectly well with taller vegetation because they have unexpectedly – some would say freakishly – long legs; if you see a hedgehog moving at speed, you'll notice that it looks like it's picking up its skirts to run. Some radiotracking studies have revealed that hedgehogs often forage around lawn margins, adjacent to longer vegetation. This makes perfect sense: ground beetle diversity and abundance are highest where there's a mixture of vegetation heights. As for the yellow underwing so heavily favoured by hedgehogs, the adult moths like dense vegetation for laying their eggs, so they tend to occur in less rigorously managed habitats.

The problem with habitat creation is that it's not a quick win. Understandably, many gardeners are more inclined to knock a hole through their fence, install a hedgehog house, or put out food overnight. (N.B. Hedgehogs are lactose-intolerant, so don't give them milk or dairy products; they can die of diarrhoea.) Feeding wildlife is big business in the UK and, despite mixed evidence on conservation benefits, it's actively encouraged by many wildlife charities, which, by happy coincidence, receive significant income from feed companies in the form of sponsorship. In the UK alone, annual expenditure on wild bird food exceeds £240 million per year. That's enough food for three times the entire breeding populations of the ten commonest feeder-visiting species, even if they ate nothing else. Supplementary feeding causes changes in community composition, favouring bold species such as blue tits at the expense of shy ones like willow tits. Now most pet shops and garden centres have started to sell what's labelled as 'hedgehog food'. A camera-trap project monitoring hedgehog feeding stations has discovered high rates of interactions between hedgehogs and other species, including badgers, foxes, rats and cats. In the ensuing argy-bargy, hedgehogs more than hold their own, often shoving bigger but less prickly species (even badgers) out of the way; they come off best in almost half of their unfriendly interactions with foxes. Whether these feeding places increase the transmission of zoonotic disease, or encourage artificially high population densities, we just don't know. We also have no idea whether supplementary feed is a nutritionally adequate substitute for a wild diet. In many cases, hedgehogs are barrelling from one suburban garden to the next, no longer bothering to eat boring old invertebrates on the way.

So keen are hedgehogs to access their fast-food takeaways that they seem unconcerned by artificial light in gardens.[6] The worry is that night lighting isn't a good thing for hedgehogs, nor for their prey. Any moth falling dead from exhaustion beside a window or streetlight isn't doing what all good moths should – feeding, having sex, laying eggs that will hatch into tasty caterpillars for birds and hedgehogs, and getting themselves eaten by bats. So there's a very simple way to help hedgehogs and other wildlife: switch off your porch lights and draw your curtains at night.

In the winter, you're more likely to find a hedgehog in a garden with supplementary food than one without.[7] Researchers comment that stimulating increased activity at this sensitive time could push hedgehogs into a net energy deficit or, conversely, help some individuals survive that might not otherwise make it through the winter. Feeding hedgehogs isn't a neutral or uncomplicatedly positive activity, so – as scientists always say when they want more funding – further research is urgently needed.

One thing that we don't want hedgehogs to encounter in gardens is pesticides, and there's good news here thanks to an unlikely hero. Accustomed to seeing little positive action from Defra on environmental issues, most conservationists were surprised by the flurry of activity that accompanied the arrival in 2017 of Michael Gove as Secretary of State for Environment, Food and Rural Affairs. Gove, we can all agree, is the sort of politician about whom it's hard not to have a strong opinion one way or (more likely) the other, but we should unanimously welcome one of his earliest actions: to introduce legislation that prohibited the sale of metaldehyde with immediate effect and outlawed its use on farms. The move was met with howls of protest from the manufacturer, supported of course by the National Farmers' Union which claimed to be concerned about the efficacy of alternative control methods. Taken to the High Court, Defra conceded that it had made procedural errors and withdrew the legislation. Armed with further advice from the UK Expert Committee on Pesticides and the Health and Safety Executive, and this time taking care to tick all the right boxes, Defra then introduced a phased ban, so that since April 2022 the use of metaldehyde anywhere in the UK has been illegal.

Before we get too excited, we need to remember that the impacts on hedgehogs from metaldehyde were likely to be low. Experiments in Germany with six captive hedgehogs that were each fed 200 metaldehyde-contaminated slugs failed to detect any adverse effects, while just three out of 370 hedgehogs found dead on roads in Norfolk in the 1970s and '80s were diagnosed with metaldehyde poisoning.[8] Hedgehogs, as we've seen, aren't keen on slugs and snails in the first place, and poisoned slugs and snails are even less delectable because they're covered in fizzed-up slime. We need now to tackle broad-spectrum insecticides. Their application can be orders of magnitude higher in domestic gardens than in agricultural fields, and it's nothing short of remarkable that you can wander into any supermarket and buy these ghastly chemicals. What a pity that Gove didn't ban them for garden use while he was at it.

Another serious risk to hedgehogs comes from anticoagulant rat poisons, and here we see no interest from the government in pushing through a ban despite the clear evidence that they're horrifically cruel. The bait is designed to be irresistible to a wide number of species, and while strict protocols established for professional contractors try to minimise the exposure to non-target animals, anyone can buy rat poison from a supermarket or agricultural store. To make matters worse, invertebrates feed on the poison with no ill effects, because they aren't susceptible to anticoagulants. They're then eaten by hedgehogs and other wildlife. We've encountered at least one terminally sick hedgehog with bleeding that indicated poisoning. Two-thirds of 120 dead hedgehogs submitted by wildlife rescue hospitals to the Centre for Ecology & Hydrology in Lancaster showed exposure to first- or second-generation anticoagulant poisons,[9] and almost a quarter of the exposed animals contained more than one type of poison. None had signs of haemorrhage, which implies that the doses were sublethal. But if nothing else, this illustrates the appalling extent of the contact between wildlife and a cocktail of highly toxic rodenticides. At the moment we don't know what the long-term effects of these exposures are; to determine the measure of a lethal dose would involve grisly experiments that nobody's in a hurry to conduct. The straightforward way of getting rid of rats while conserving your hedgehogs is to prevent access to water, waste food, compost bins and chicken feeders. If you must call in the professionals, insist they don't use poisons.

Hedgehogs have something in common with sheep: they're experts at finding ingenious ways to die. Our community noticeboard here in East Devon implores footballers to raise goal nets off the ground when the game's over so that hedgehogs don't tangle themselves up. McDonald's changed the size of its lids because hedgehogs, intent on guzzling the remnants of McFlurry ice cream, kept wedging their heads in the cups. They get frazzled by electric fences, suffocated by plastic bags, stuck in drains, burned in bonfires, jammed in cattle grids, and drowned in hot tubs.

Hedgehogs are also vulnerable to a staggering array of infections. The Garden Wildlife Health Project, a collaborative effort run from London Zoo, has discovered all sorts of parasites and pathogens in the hundreds of dead hedgehogs submitted by the public. One of the most ubiquitous is lungworm – or lung*worms*, as there are actually two nematode parasites to blame: *Crenosoma striatum* and *Capillaria aerophila*. Not even their mothers love those little wrigglers. The eggs and larvae are excreted by hedgehogs in their droppings. Once in the environment, they infect slugs and snails (for *Crenosoma striatum*) or earthworms (for *Capillaria aerophila*), which are then eaten by other hedgehogs, and so on. A depressingly high proportion of hedgehogs in Britain and Europe are affected by lungworm; Fiona's own kitchen-table post-mortems of hedgehogs we've found dead or dying often expose lungworm as the culprit. Its impacts vary depending on how many worms are present and whether there are associated bacterial infections, so symptoms can range from a mild snuffle, to severe wheezing and loss of appetite, to death. Whether we should be proactively treating hedgehogs for lungworm and other parasites, as we do our pets, is a trickier issue. Natural immune responses to infection usually keep parasite burdens to manageable levels in wild animals that are otherwise healthy. Anti-parasitics could therefore just be treating the symptom rather than the cause, and potentially risk compromising the immune system in the process.

Normally, we think of animal welfare and conservation as two distinct priorities: welfare focuses on individuals while conservation takes a community- or species-level perspective. In the case of British hedgehogs, we're fast approaching the point where these strategies need to merge. Something like 40,000 hedgehogs pass through hogspitals every year, which is getting close to five percent of the total number of wild hedgehogs in Britain. Those centres have reached capacity and are turning away casualties: we found our latest injured hedgehog in Hampshire as we were making our way from Devon to Norfolk, and not a single one of the ten rescue centres en route had space available. Against this surge in demand, there's only so much that organisations dependent on public donations and the extraordinary dedication of volunteers can achieve. The government needn't stand idly by. It could start by subsidising the vets' bills for patching up hedgehogs. It could also test the effectiveness of projects like the Brackley Hogwatch, where teams of volunteers went out and weighed as many hedgehogs as they could find, taking in for further attention any that were underweight or unwell. If results show that these campaigns have a significant effect on numbers at a local level, a national programme may yet save the hedgehog from terminal decline.

Enough with the gloom and despondency. When it comes to quirky facts, we don't want Rory Stewart to hog the limelight, so here goes:

The Japanese word for hedgehog, *harinezumi*, transliterates as 'needle mouse'. Hedgehogs are mice? Ridiculous! As every English-speaker knows, hedgehogs are pigs that live under field boundaries. That's why males are 'boars' and females are 'sows'. The great Swedish taxonomist Carl Linnaeus seemed to agree: he grouped hedgehogs with pigs for no better reason than that they were both 'long-snouted'.

Hedgehogs – hold the front page! – are neither mice nor pigs. Nor are they related to other spiny mammals like porcupines and echidnas; any resemblance is merely the result of convergent evolution. Some tenrecs, a family of mammals endemic to Madagascar, appear so strikingly similar to hedgehogs that they used to be included in the same taxonomic group: the order Eulipotyphla, which also includes moles and shrews, and rudely translates as 'truly blind and fat'. Around twenty years ago, genetic analysis showed that tenrecs weren't close relatives at all, at which point they were unceremoniously yanked out of the London Natural History Museum's display case and shunted alongside elephants and aardvarks in the Afrotheria. There's an ongoing campaign to give hedgehogs their own order; after all, they diverged from shrews and moles between 65 and 70 million years ago.

Hedgehog taxonomy amounts to a huge bunfight, which is why we're being deliberately vague when we say that there are something in the region of fourteen species of hedgehog across the Old World. (Wikipedia currently lists seventeen, while debate in the scientific literature revolves, as ever, around where you think that species stop and subspecies start.) You won't find hedgehogs in the Americas, and you wouldn't find them in Australasia had some bright spark not introduced them to New Zealand in the nineteenth century.

Our UK brand of hedgehog is *Erinaceus europaeus* – the largest of all the hedgehog species – with a range across western Europe and much of Scandinavia. Minor divergences between the British model and its continental brethren have started a Hedgexit debate about whether it should be considered a separate subspecies.

Every schoolboy knows how hedgehogs have sex: very carefully. Aristotle, wrestling with the same conundrum, concluded that they must mate belly-to-belly. He was wrong and the schoolboys were right. The female arches her back and flattens her spines, and the male, having carried out a quick cost–benefit analysis, clambers painfully into position, sometimes grabbing hold of her shoulder spines with his teeth to aid stability.

To check the sex and breeding status of a hedgehog – making sure, for example, that it's not a lactating female – the quickest and most effective way is to sit the animal on the lid of a glass casserole dish and take a peek from below. This does actually work. Try instead to turn a hedgehog over and it'll roll into a ball. It can stay like that for hours; you'll definitely get bored before it does.

In the same way that sex can be a risky business for amorous hedgehogs, giving birth sounds eye-wateringly painful. The possibility that the hoglets might emerge tail first – against the grain, as it were – doesn't bear thinking about. Fortunately, hedgehog mums are spared thanks to a temporary adaptation: the hoglets' spines are submerged in skin tissue inflated with fluid. The fluid drains away during the first day after birth.

The Ancient Greek poet Archilochus is credited with the observation that 'The fox knows many things; the hedgehog, one big thing.' The implication is that a fox is cunning and resourceful, whereas the hedgehog's single trick of rolling into a ball is enough to frustrate most predators – foxes included. Isaiah Berlin used this distinction to separate and classify thinkers such as Plato (a hedgehog because he interprets everything through a single big idea) and Aristotle (a fox because he believes that no single idea can do justice to the variety of existence).

When hedgehogs aren't sucking on cows' teats, they're making off with stolen fruit on their spines. That, at least, was the view of Pliny the Elder, writing in the first century CE, and it was frequently repeated by medieval authors. There are wonderful images in bestiaries of hedgehogs adorned with grapes and other fruits so that they can take them back to their offspring. The myth hasn't quite died out. Some people swear that they've seen hedgehogs carrying fruit, even though experiments to push grapes onto hedgehog spines and make them stay there have been comically unsuccessful.

In the hierarchy of tall tales, the fruit-impaled-on-spines story falls short of the belief that hedgehogs cause crop circles. Even while dismissing this as nonsense, we have to concede that the suggestion has a certain creative logic. Circle-running is a rare phenomenon among hedgehogs, but we've seen it several times ourselves; the hedgehog will go around and around for no apparent reason and with no sign that it's ever going to stop. Some experts suspect that the behaviour is caused by neurological damage; bacterial or viral infection may also play a part. Regardless, crop circles are made by flying saucers.

One final hedgehog myth has just a bit more substance: it's true that, in the Victorian period, hedgehogs were sometimes kept in kitchens and pantries as roving pest-control devices. Beatrix Potter's pet hedgehog may have performed that function for her owner. We can't recommend it. If you have hedgehogs in your garden, you'll notice that their droppings are deposited haphazardly across the lawn, not grouped in a latrine. House-training is anathema to hedgehogs. Quaint though the idea of a pantry-hog may be, food hygiene inspectors would take a dim view.

Self-anointing is the strangest behaviour of this strangest of mammals. What it involves can be guessed from a German description coined in 1912: *Selbstbespuchen*, which is routinely translated as 'self-spitting' and more exactly means 'spitting *on* oneself'. Without warning, the hedgehog will start to foam at the mouth and contort oddly, using its long tongue to deposit generous lashings of saliva around its body and especially its spines. The hedgehog shows no awareness of surroundings; it enters what seems to be a state of extreme mental abstraction, sometimes for several minutes and even hours, while it focuses completely on the task at hand. It emerges from this trancelike state dazed and confused. The behaviour is common to all hedgehog species, and to a small number of other mammals, such as – believe it or not – the giant panda. Self-anointing is harmless and perhaps beneficial, but it's been known to send owners of African pygmy hedgehogs into a wild panic as they assume that their beloved pet is having fits and dying before their eyes.

There are countless theories about what's going on; well, we've counted seven. Common to the majority of them is the assumption that self-anointing has something to do with scent, because usually one of the triggers is a strong or new smell, such as a nearby object made of leather. So if you put a watch strap next to a hedgehog, it may start by sniffing it, then it'll lick and chew, before frothing at the mouth and flinging that lovely leather-tainted saliva over its spines. (On the other hand, it may just ignore the watch strap altogether.) Nigel Reeve lists well over thirty objects that have been known to induce self-anointing, many of them distinctly pungent (perfumes, tobacco, creosote), but others utterly bizarre (toads, nylon stockings, woollen carpets).[10] The drug of choice for one of our rehabilitating hedgehogs was wood, particularly the wooden floor in our spare room, which she licked with such ardour that we could barely make out her nosetip through the foam.

Something Beatrix Potter forgot to mention is that, given the chance, hedgehogs will enthusiastically coat themselves in dog poo; there's more than one reason for wearing gloves if you should ever need to pick one up. Dog poo is also a favourite stimulus for self-anointing, which fits the assumption that the hedgehog wants to smell like whatever the unusual object happens to be. They're naturally quite smelly creatures even without external assistance, so the 'camouflage hypothesis' argues that they're deliberately masking their own scent in order to avoid the attention of predators and bigger hedgehogs. This is supported by evidence that, although self-anointing takes place among males and females and among adults and juveniles, the group most likely to indulge is the subadult males. They have good reason for wanting to move around under the radar, even if stinking to high heaven of a foreign substance seems an unlikely way of going about it. The problem with the hypothesis is that in practice it doesn't work. There's no evidence that self-anointed hedgehogs are more likely to avoid predation, and in fact sniffer dogs find them just as easily as ones that haven't self-anointed. Accounts from hedgehog carers tally with our own experience that hedgehogs smell more rather than less – to use the technical term – *hedgehoggy* after bouts of self-anointing.

Perhaps, then, the purpose of self-anointing is to enhance and individualise scent rather than to camouflage it. Dogs don't self-anoint, but they'll sometimes roll in smelly substances just to advertise their presence and feel good about themselves: our own dog has a penchant for the cloying jasmine scent of otter spraint, and once rolled thoroughly, not to say exhaustively, in a dead and deliquescent fox. Something similar may be going on with hedgehogs when they self-anoint: they broadcast their own natural fragrance together with base notes of a certain enigmatic *je ne sais quoi*. This has a number of benefits, such as to mark territories, to send signals to potential mates, to warn off rivals, or to forge bonds between mothers and their offspring: hoglets have been observed self-anointing with their mother's milk. It seems that hedgehogs self-anoint to create what Nigel Reeve has called 'a general-purpose personal odour'.

*

Pricole Kidman was our hedgehog with the wood fetish. Tim had found her tottering outside Plymouth Argyle's stadium – home of English football – early one Sunday morning. She'd suffered a head injury, probably having been hit by a car. We assumed that she wouldn't survive, and it soon became obvious that she was blind or very nearly. Somehow she recovered and, although she'd never be released, we were assured by our wildlife vet that she could still enjoy a good quality of life. Then for seven wonderful years she thrived; she liked nothing better than to sit on our laps and be stroked, for which she'd graciously relax her spines. (You can soon tell when a hedgehog isn't happy: it snorts, tucks its head in, and pulls its spines upwards and outwards to make a crown of thorns.) Pricole ended up as something of a celebrity, with appearances on the BBC's *Spotlight*, *Channel 4 News* and even several radio programmes, whenever Fiona had a story to tell about declining hedgehog populations. She modelled for the front cover of *Mammal News*, although on that occasion a touch of stage fright meant that she was photographed rolled up in a ball. Pound for pound, she had the biggest yawn of any creature ever. She was the best hedgehog and we loved her.

It's important to establish our credentials because of what we have to say next: hedgehogs are killers. Of course that's obvious if you're a beetle or a caterpillar, but the slaughter doesn't stop there. They'll eat adders and seem to have a partial immunity to venom; they grab hold of the snake somewhere along its body, bite through until they reach the spine, and gradually work their way up towards the head. It's not a quick death. Hedgehogs do something similar when they come across bird nestlings, and there are many anecdotes of them attacking fully mature birds and even a Canada goose. We think of spines as a defence mechanism, but they also transform hedgehogs into juggernauts. Maybe they're not so reserved after all. Their singlemindedness in pursuit of a meal may well explain the myth that they steal milk from cows. Their mouths aren't big enough to manage that – and how would they know what an udder was for? Yet there are some old and semi-reliable accounts of hedgehogs determinedly hanging onto the udders of cows that had made the mistake of lying down in a field. Perhaps from certain perspectives an udder looks like a slimeless slug.

Despite a weakness for suitably sized eggs, hedgehogs usually come well down the list of threats to ground-nesting birds, behind raptors, corvids, gulls, cats, badgers, dogs, weasels, stoats, rats and, of course, humans with their industrialised farming methods. Dump hedgehogs on a small island where they're not native and leave them to breed unchecked alongside colonies of waders and different rules apply. On North Ronaldsay during the early 1970s, a pair of hedgehogs was introduced in the hope that it would keep down slugs in someone's greenhouse; this erinaceous Adam and Eve generated more than 500 individuals within a decade. The same processes turned four hedgehogs into several thousand on South Uist over thirty years, and obliged wildlife enthusiasts to pick sides. After hearing about a proposed hedgehog cull on Uist, guitar-hero-slash-environmental-campaigner Brian May declared himself 'outraged at the logical absurdity of a bunch of birds of very small brains being put above these delightful and intelligent mammals'. We're Team Mammal too, although we're not sure that intelligence should be the criterion for deciding what to conserve. Having spent several decades looking after waifs and strays, we can promise that hedgehogs are the prickliest but not the sharpest tools in the shed. Anyone who's ever lost an ice cream to a herring gull careering out of the midday sun will testify to an extreme avian cunning that gives 'bird-brained' an entirely new meaning. Gulls are effortlessly capable of outwitting humans. Nobody has ever been outwitted by a hedgehog.

The story – more accurately the fiasco – of Ronaldsay and Uist has been told before, and we recommend Hugh Warwick's brilliant account in *A Prickly Affair* to anyone wanting the forensic details.[11] The pantomime villain of the piece was NatureScot (at the time called Scottish Natural Heritage), which in 2003 pushed through its culling policy on Uist with no evidence that it could entirely eradicate the hedgehogs. This came at vast expense to the taxpayer: estimates fall somewhere between £350 and £1,000 per dead hedgehog. Strangest of all, NatureScot had ignored a great deal of existing evidence in order to convince itself that translocating hedgehogs to the mainland was far crueller than a lethal injection. Leaving aside ethics, economics, logistics, science and PR, the cull was a great idea. At one point, rival bands of cullers and rescuers were patrolling the same small island every night, racing each other to gather up hedgehogs whenever they encountered them. Warwick reports that, after four years of culling, the rescuers were ahead by 756 to 658. Meanwhile, the seabirds still weren't breeding. They weren't breeding especially well on hedgehog-free islands, either. After all the noise had died down, very little research was done to monitor the long-term impact of hedgehog removal, but it does seem that another unrelated cause of the breeding colonies' collapse had been a sharp decline in the availability of sand eels.

All sides should be careful when assessing the damage to bird populations by hedgehogs. Some ornithologists sound like those deluded spokespeople from the Angling Trust who seem to fret far more about the reintroduction of beavers than the 400,000 incidences of raw sewage being discharged into our rivers every year. At a time of climate change, overfishing and the continued polluting of our seas, the overall threat from hedgehogs within their *native* range is minuscule. Then there's the issue of distinguishing hedgehog damage from the far greater destruction wrought by cats and dogs. Should we be culling them, too? Even so, mammal biologists need to make concessions wherever hedgehogs are an invasive species; their presence may prove to be the tipping point for breeding colonies already struggling under adverse conditions. At the height of the hedgehog problem on Uist, you could take a torch onto the machair – grassy plains just inland of the dunes – and count them by the dozen. The machair provides one of the most important breeding sites for wading birds in Europe. Some species of gull stand no nonsense, but waders such as lapwings are polite to a fault, and will shuffle to one side while the hedgehog butts in and snacks on their chicks and eggs. NatureScot now blames hedgehogs for 'up to fifty percent' of all breeding failures on South Uist among lapwing, dunlin, ringed plover, redshank and snipe. 'Up to fifty percent' is a transparent attempt to give as high a number as possible while staying deniably vague, but NatureScot did film many of these hedgehog raids and posted them on social media. There are more inspirational things to watch, so we'll take their word for it.

You can look at a much bigger canvas than the Scottish islands for proof of hedgehog damage to biodiversity. Hedgehogs were introduced into New Zealand for the first time in 1869, and they've been busily crunching their way through native invertebrates and reptiles ever since; they're also implicated in the catastrophic decline of at least a dozen bird species. There are now far more hedgehogs in New Zealand than in the UK. These antipodean larrikins are seriously bad news, and vast amounts of money are being thrown at an eradication project.

As the experiences in New Zealand and the Scottish islands show, hedgehogs have an uncanny knack of flourishing only in the wrong places. For obvious reasons, they like locations with no natural predators and low traffic levels. You can currently find what must surely be the highest density of hedgehogs anywhere in Britain on St Mary's, the largest of the islands that make up the Isles of Scilly. They're non-native and were introduced in the 1980s when, so legend has it, three were brought over from the mainland as a birthday present for a young girl. All escaped, but, while one was found dead the next day, and a second somehow contrived to get itself killed on a quiet road within a week, the survivor was (inevitably) a pregnant female from whom all hedgehogs on St Mary's since are descended.

It's a lovely idea that we can solve an extinction crisis by growing what's been called an 'insurance population' in a protected space, but history shows that the losses to native flora and fauna – themselves often endangered – far outweigh any gains. We need only ponder the case of the Tasmanian devil, a winsome but homicidal marsupial that sinks its teeth into anything made out of meat, including other Tasmanian devils. In the mid-1990s, the species started to succumb to a contagious facial tumour spread by their unfortunate tendency to bite chunks out of each other. With numbers freefalling, twenty-eight healthy individuals were rounded up and relocated from the Tasmanian mainland (where they're native) to the remote Maria Island (where they aren't). Ten years later, there were close to a hundred little devils scampering around the island, dining on an exotic melange of possums, wombats and, most of all, seabirds. To the surprise only of the conservationists who devised this cunning plan, during the same period the island's breeding pairs of penguins shrank from 3,000 to zero, and the shearwater colonies disappeared without trace. Shearwaters nest in burrows; handily, Tasmanian devils are really good at digging.

Hedgehogs obviously aren't having that level of impact on St Mary's, although their introduction did – we're choosing our words carefully – *precede* the disappearance of skylarks on the island. Fiona very nearly encountered a hedgehog on St Mary's once before, when in 2011 she went on a busman's holiday with friends from Wiltshire Bat Group who were trying to identify which bat species were present on the Isles of Scilly. One night, they caught and radio-tagged a brown long-eared bat – the first record for the species anywhere across the archipelago for more than forty years. They spent much of the next day trying to locate its roost, only to discover when they returned that a sneak thief had nosed into one of the tents and started to munch through a packet of chocolate biscuits. The guilty party had added insult to injury by leaving its calling card: there, atop a sleeping bag, lay a squishy, black, tapered, slug-shaped and therefore utterly diagnostic poo.

We don't need much of an excuse to visit the Isles of Scilly, so we decided that in the best interests of science we simply had to go back and assess what impact the hedgehogs were having on St Mary's. Has their population density levelled off? Are they damaging local biodiversity? If so, can we just scoop them up and hedgehog-bomb them across the parts of Britain where they're so urgently needed? Questions, questions – and the only way to answer them was to sail away to St Mary's on a hedgehog hunt.

The Isles of Scilly may be a paradise teeming with seabirds, butterflies, beautiful flora and the occasional stray walrus – more about him in a moment – but when it comes to terrestrial mammals the islands have a history of more-or-less disastrous human interventions. In 1855, the Lord Proprietor cleared Samson of its human inhabitants, who had been wasting away on a diet of limpets and potatoes. In their place, he established a deer park; so ungrateful were the deer that almost all of them drowned in their determination to escape, although a small number are reported to have survived the kilometre swim to Tresco. The same man reputedly introduced colour-coded rabbits to the Isles, a different colour for each island; today they're all black or brown, and seem to have disappeared from some islands altogether. Brown rats arrived from ships (or, if you prefer the more romantic version, ship*wrecks*) at some point in the eighteenth century, and continue to have a catastrophic impact on seabird populations; rats have now been eradicated at great expense from Annet, Menawethan, Gugh and St Agnes (where daytrippers are greeted with a large sign reading 'Rat on a Rat' and a phone number to report any sightings), but they persist on several of the larger islands. Hedgehogs aren't even the latest of the non-native species: red squirrels were introduced to Tresco in 2012, although, inconveniently, the majority of them soon died of septicaemia and injuries following fights. The survivors were supplemented the following year when a delivery arrived by Royal Navy helicopter from a Surrey zoo, since when their numbers have increased more than fivefold. This is described by collaborators as a project to help save red squirrels in Britain by creating a viable offshore colony – which overlooks the existence of thriving populations across Europe.

A much more compelling reason for visiting the Isles of Scilly is to find the lesser white-toothed shrew, more commonly known as the Scilly shrew because it doesn't occur anywhere else in Britain and no other shrew species occurs on the Isles of Scilly. This splendid isolation was briefly challenged in 2010, when an individual shrew made national headlines by stowing away on a ferry to Penzance, only to be flown safely back home to St Mary's the following day. You can identify a lesser white-toothed shrew by the fact that all other British species have red-tipped teeth, but suffice it to say that if you're able to inspect a shrew's teeth that closely, it's dead; with a live one, even the shrewdest of shrew experts may struggle. Your best bet is to look for a triangular head and a conspicuously long, almost trunk-like snout. The simplest way of all is to ask yourself where you are: if the answer is the Isles of Scilly, what you have before you is a Scilly shrew, and if not, it isn't.

Quite how and when the lesser white-toothed shrew turned up on the Isles of Scilly is anyone's guess. It's possibly yet another species introduced through human activity, but that must have happened in the distant past. Although archaeological evidence for such a small species is sparse, we know that it had already arrived 3,000 years ago. 'Arrived' may be the wrong word; perhaps it was there all along. A land bridge joined Cornwall to the Isles of Scilly during the last Ice Age: this was the semi-mythical Lyonesse, and if you buy a fisherman a drink, he'll regale you with stories of how on a still day at sea you can hear the bells of drowned churches faintly ringing in the ocean currents. All that landmass between Cornwall and the Isles was inundated as the ice melted (quite a few thousand years before Christianity), but land bridges between several of the islands have only finally been covered in recent centuries. The lesser white-toothed shrew may be a 'glacial relict' – a sole survivor stranded on the Isles of Scilly even as it was outcompeted by other shrew species on the mainland. Or it may simply have hitched a ride from what is now Spain or France on a Bronze Age boat. These debates explain why the lesser white-toothed shrew is given a keeping-all-options-open classification in Fiona's population review: 'Non-native (naturalised), but possibly native.' The Red List errs on the side of caution and treats it as a native species, with all the attendant legal protections: currently, none.

Why this detour into shrew territory amidst a chapter on hedgehogs? Because we have no idea what effect the hedgehogs are having on them. Shrews nest underground, away from greedy paws, but they compete for the same limited food resources: beetles and grubs. Whereas the waders on Uist had powerful friends – the RSPB's annual income exceeds £100 million – nobody bothers about shrews. It isn't even possible to come up with a sensible stab at how many there are; the unreliable hedgehog guesstimates seem finely calibrated by comparison. The Scilly shrew is classified as Near Threatened on the Red List; before we can move it to the more serious category of Vulnerable, we need to know that population size, range or habitat quality is declining. At the moment, we don't have information for any of these things. If the hedgehogs on St Mary's turn out to be a problem, the shrews may disappear long before anybody notices.

The gods are smiling on you when you sail into St Mary's hoping to find hedgehogs and shrews, only to dock near a pontoon on which lies sprawled – of all things – a 750-kilogram Atlantic walrus. Wally (as he'd been christened) was the handsomest creature we'd ever seen. He'd been in residence for the past few weeks, and worked to a consistent schedule: he'd lounge around for a couple of days without a care in the world, before heading to one of the smaller islands to gorge on crustaceans. Just when the locals on St Mary's began to assume that he wasn't coming back, he'd turn up slap bang in the middle of the busiest harbour for forty miles and make himself at home once more. Some people insisted that he was stressed, but Wally was clearly having a walrus of a time, and not even a 1,200-tonne ferry chugging along eighty metres away was enough to disconcert him.

We landed on St Mary's having already seen Wally, a pod of common dolphins and a playful pair of harbour porpoises. Shrews were likely to prove more elusive, but we'd brought our secret weapon: a box of twenty Longworth traps. For the mammal ecologist, these devices are what a hammer is for a carpenter. You can't do the job without them. They comprise a lightweight aluminium chamber and a smaller tunnel that juts out from it. The mammal enters the tunnel, lured by bait or natural curiosity, and trips a mechanism so that a door closes behind it; the creature then scuttles along to the larger chamber where bedding and a ready meal await, until it's identified and released.

Longworth traps are perfect for surveying mice and voles, and Fiona once accidentally caught a weasel in one; she knew something had gone awry because the trap was clattering up and down against the ground. Awkwardly, there's a particular problem with shrews, which is why you need a special licence to trap them. With a heart rate of more than sixteen beats per second, their high metabolism necessitates a frantic race to find and eat ninety percent of their body weight every day. This means that they have an alarming tendency to topple over dead if held in a Longworth trap for any length of time. The solution under normal circumstances is to buy traps with specially designed 'shrew holes' that are too small for voles and mice but allow shrews to escape. That's not much good if it's shrews that you happen to be trying to catch, so *faute de mieux* we added a banquet of different foods and checked the traps every couple of hours. We're relieved to report that no animals were harmed in the making of this chapter.

At least, not by us. While we were erecting our tent, we heard a high-pitched scream about thirty metres away. That was our first shrew, disappearing down a gull's throat. Later in the evening, having just put out the Longworth traps, we noticed that the door to one of them was already shut. This isn't especially unusual, because setting traps is fiddly and they can close without the slightest provocation. So Fiona reopened the trap, only to be startled when out sprang a small mammal, up her left arm, across her neck, down her right arm and away. That was our second shrew, although it was more blur than mammal. What we thought was going to be our third shrew turned out instead to be a beautiful wood mouse. Over the next three days we caught a grand total of two more shrews, both of them in grassland and none in bracken. That's a lot of trapping for not much reward, although with such small sample sizes it's impossible to draw conclusions about population numbers.

As for the hedgehogs, well, they were everywhere. In the areas of coastal heath where we were trapping shrews, the challenge wasn't so much finding them as avoiding tripping over them. St Mary's is the least important island on the archipelago for wildlife, but if hedgehogs reached any of the others there'd be an urgent issue for seabirds. In the meantime, before getting too exercised about the impact of hedgehogs on shrew populations on St Mary's we'd want to look at the prevalence of rats, as well as doing something about the feral cat and her two kittens that were exploring the foreshore just down the hill from where we were working. Let the hedgehogs have their island utopia, while their mainland kin take their chances with badgers and cars. We don't even begrudge them our food. During our last night, a particularly determined hedgehog chewed through two plastic bags to gorge himself on a slab of butter we'd stored by our tent. We caught him in flagrante (which was just as well because YOU MUST NEVER FEED DAIRY PRODUCTS TO HEDGEHOGS!), but he just snorted disdainfully and shuffled back into the shadows.

SEVEN

Who Cares What Colour the Squirrels Are?

It may sound harmless, but that question's the fastest way to start a war between animal rights campaigners and conservationists. The conservationists focus on the long-term survival of the native red squirrel, and are prepared to sacrifice any number of invasive grey squirrels to achieve it; the animal rights campaigners deplore the culling and insist that the ends don't justify the means. What makes matters worse is that both sides are losing. As red squirrels gradually disappear from mainland Britain, thousands of greys are being killed in a rearguard effort to save them.

One reason for the falling-out is that longstanding labels like 'native', 'non-native' and 'invasive' seem to echo public debates around immigration. We've sometimes heard it insinuated that – to put it crudely – if you're hostile to grey squirrels you're a bit racist. In scientific terms, the comparison makes no sense. Whether Inuit or Indigenous Australian, humans are a single species so closely related that it's impossible to divide us into subspecies. Red and grey squirrels, on the other hand, are separate species that aren't able to interbreed or flourish side by side. Grey squirrels destroy reds by unwittingly conducting biological warfare against them. Many greys carry squirrelpox virus, which usually does them no harm at all but is fatal for any reds that they encounter. Take your pick: if you don't cull greys, you're condemning the reds, and death from squirrelpox isn't pretty.

Over the past fifty years, governments worldwide have spent an estimated $1.3 trillion trying to control non-native species that devastate local biodiversity.[1] Conservationists worry about what they sometimes call the McDonaldization of ecosystems, whereby a small number of species capable of thriving alongside humans will spread across the globe, outmuscling native flora and fauna and imposing homogeneity. Invasive species are one of the primary causes of biodiversity loss, and waging war on these colonisers can seem more appealing than tackling the diffuse issues of habitat destruction and human population growth. Yet so much of our British wildlife is already non-native that it's absurd to try to return to some factitious state of absolute innocence. What about rabbits, little owls, Canada geese, sycamores and horse chestnut trees? For many children delighted by their antics in parks and gardens, grey squirrels are the gateway drug to a lifetime of positive engagement with the natural world. They're every bit as intelligent and characterful as the reds. We'd do well to bear that in mind next time we're trying to work out how many dead greys a red squirrel is worth.

In 1876, a banker by the name of Thomas Brocklehurst arrived home from a business trip to the United States with a pair of Eastern gray squirrels. The story goes that he kept them in a cage to show friends, but released them into his estate at Henbury Park in Cheshire when the novelty wore off. If, having throttled baby Hitler in his cot, we could travel further back in our time machine to terminate those squirrels before they bred, it wouldn't make the slightest difference. Brocklehurst probably wasn't the first to introduce grey squirrels, and he certainly wasn't the last. There were another thirty releases or more across the British Isles over the following half-century, including introductions in Wales and Ireland sourced from a thriving new population of greys at Woburn. Sometimes as many as a hundred individuals were freed at once, so it's no surprise that the greys became established in their new landscapes.

In 1932, as if testing the principle 'better late than never' to breaking point, the government made it illegal to release grey squirrels anywhere in Britain. Tighter regulations in recent years mean that if a grey gets trapped in your shed and you open the door to let it out, you're breaking the law, as is a wildlife rescue centre returning a rehabilitated grey squirrel to the wild. Unbothered by these legal niceties, greys have continued to spread relentlessly from their various introduction sites, pushing reds to the margins. Islands have long since become the most secure and defensible refuges: there are colonies of reds on Anglesey, the Isle of Wight and Brownsea Island. Don't bother looking for reds across the south of mainland Britain.

Things are marginally less depressing further north, where intensive campaigns to create a Maginot line and halt the greys' advance have proven partially effective. Scotland remains a stronghold for red squirrels – but not, alas, an impregnable one. The 2018 population review makes an estimate of 239,000 red squirrels in Scotland (where they're Near Threatened), while they're Endangered with 38,900 individuals in England and

9,000 in Wales. Since the mid-1990s, reds have disappeared from East Anglia, the Humber Estuary, Derbyshire, Pembrokeshire, Carmarthenshire and Denbighshire. Meanwhile, grey squirrels have been busily expanding their range across mainland Britain, where the 2.7 million greys outnumber reds by about twelve to one and rising. Worryingly, 478,000 of those greys have been spreading north of the border. Huge efforts have halted their expansion beyond the Central Belt, so their distribution in Scotland remains about the same as it was in 1992. Nevertheless, grey squirrels continue to turn up in the unlikeliest of places: one reached the Isle of Skye, having hitched a ride from Glasgow under a car bonnet. The first case of squirrelpox virus among reds in Scotland was reported in 2007, and now it's commonplace.

In England, a fifty-hectare forest has been holding out heroically against this invasion. The smallish town of Formby, on the west coast twelve miles north of Liverpool, is home to what is probably the most studied red squirrel population in the world. (Its main rival is the colony on Anglesey, thirty miles away if you're swimming it.) Formby encapsulates the squirrel wars in microcosm. Besieged on three sides by greys, it survives as a haven thanks to the sustained efforts of the National Trust, wildlife organisations and a team of committed volunteers. Even with extensive operations to trap and cull greys in the area, there have been catastrophic outbreaks of squirrelpox virus, and numbers of reds are still nowhere near the levels seen in autumn 2007. Over the eighteen months to spring 2009 the population crashed by eighty-seven percent, and post-mortems revealed that most of the dead animals were infected with squirrelpox.[2] At the time, there was a strong suspicion that someone had deliberately dumped greys into the middle of the red squirrel colonies, presumably to wipe out the reds and remove the need for further culling. A smaller outbreak occurred in 2019, and then in 2021 four more diseased squirrels were found. Reds tend to die in their dreys, so the real mortality rate is always far higher than official numbers would suggest. For Formby's red squirrel population of roughly 250 individuals, apocalypse is always just round the corner.

The reds couldn't have chosen a swankier place to make their last stand. Formby boasts some of the most expensive residential properties on Merseyside, which is why only slightly more squirrels than premiership footballers live there. The appeal to millionaires is obvious: the town is posh and leafy and secluded, and a fantastic beach runs alongside the woodland. We, of course, are dedicated professionals, so it goes without saying that when we showed up on a hot summer's day, it was strictly for the wildlife.

The red squirrel has a native range reaching from the north of Portugal to the east of Russia, and further beyond to Hokkaido, the northernmost of the Japanese islands. Its scientific name is *Sciurus vulgaris* (literally, common squirrel), and, true to that description, it remains common enough to be filed away by the IUCN in the category of Least Concern. The British Isles provides the ignominious exception to the good news.

Traditionally, red squirrels are hunted for their fur and meat and because they damage trees. The Finnish word for money, *raha*, is derived from the proto-Germanic *skrahā*, meaning 'squirrel skin' – an etymology that remembers a pre-monetary system when furs served as currency. Well into the twentieth century, squirrels remained crucial to the Finnish economy, and in some years more than two million of them were slaughtered. Every so often, the government would announce a year-long moratorium to allow numbers to bounce back, but generally the squirrel population seemed resilient even in the face of these massive hunting pressures.

Consider, then, what an extraordinarily thorough job we Brits must have done to annihilate entire populations of a species found to be so robust and copious everywhere else across its range. We've been uniquely inhospitable. The arrival of the greys was merely the latest – and not even the most serious – in a long line of disasters for the red squirrel in Britain. Its ancestors had crossed a land bridge from continental Europe about 10,000 years ago; they should have stayed put. By the Iron Age, half of our original wildwood had been cleared for agriculture, producing barren landscapes that we're still encouraged to think of as natural and beautiful. (Rewilders despairingly point out that there's more tree cover in the city of Sheffield than in the Lake District.) Woodland covered only five percent of Britain by the early 1900s, and the little that survived was fragmented into small blocks. Needless to say, none of this created the ideal circumstances for a tree-dwelling rodent to prosper.

The Forestry Commission was founded in 1919 with the remit of re-establishing Britain's timber supply. Before the First World War, around ninety percent of our timber had been imported, and wartime shortages prompted the government to act. The pace picked up after 1945, with afforestation strategies that achieved the aim of increasing timber supplies but did more harm than good for biodiversity: endless rows of non-native conifers were planted, many on the sites of private woodlands that had been bought up and cleared. Simultaneously, the agricultural revolution incentivised farmers to rip out woods and hedgerows. Oliver Rackham, possibly our greatest historian of the countryside, estimated that the losses of ancient woodland in the forty years between 1935 and 1975 equalled, and perhaps exceeded, the total losses incurred during the previous thousand years.

Despite recent increases in woodland cover, Britain still has some of the lowest proportions in Europe. None of it is protected by statute for its soil, water-storage or other forest ecosystem functions; the comparable amount in Germany is forty percent. The structure of our woodland differs radically from our European neighbours'. Most is plantation (eighty-nine percent), far outstripping any other country except Ireland (eighty-six percent), and the bulk of that is non-native conifer. Only about seven percent of our woodland cover comprises mixed-aged stands, compared with twenty-five percent in Croatia. You'd reasonably expect that, having prioritised efficiency over biodiversity, the forestry sector would be crucial to our Gross Domestic Product; in fact, it contributes less than 0.5% of national GDP, which puts us almost at the bottom of the European league tables. Croatia outperforms us more than threefold, and Sweden fivefold. True, we've managed to hang on to more ancient oaks than our neighbours. But these veterans are usually isolated and out of the reach of red squirrels. We've smashed our landscapes for not very much.

It's no surprise that the decline of the red squirrel can be mapped neatly onto the decline of our woodlands. Red squirrels had disappeared from vast swathes of Scotland by the eighteenth century. Derek Yalden has charted the long inventory of loss: the last red in Sutherland was recorded in 1630, in Moray 1775, in Dumbarton 1776, and in Ross and Cromarty 1792. They'd vanished from Angus by 1813, from Argyll by 1842, and from Aberdeenshire by 1843. Finally, someone recognised the problem and did something about it. At least ten reintroductions and augmentations took place in Scotland during the hundred years after 1772, largely on the whim of landowning aristocrats who used squirrels sourced from England and Scandinavia.

From our modern perspective, it's hard to appreciate the extent of the animosity directed at red squirrels. As soon as they re-established themselves in Scotland, they were once again treated as pests. Complaints grew to a crescendo in the early 1900s, with even naturalists like James Ritchie describing the red squirrel as economically disastrous because of the damage it caused to trees. Many naturalists were also concerned about squirrels destroying wild bird nests, and they attributed the disappearance of the great spotted woodpecker in Scotland to those rotten rodents. The Highland Squirrel Club, operating in areas where the species had been extinct two or three decades previously, killed over 100,000 red squirrels between 1903 and 1946. Tails were submitted as proof, and more than £1,500 was paid in bounties. The record for the most squirrels culled went to the Lovat Estate in the Highlands – the very place where, a century earlier in 1844, Lady Lovat had persuaded the government to undertake a reintroduction scheme.

All the evidence supports the assumption that red squirrel numbers fell sharply amidst this brutal persecution. Culling stopped in the New Forest because there were hardly any squirrels left, and records for the Highland Squirrel Club show that fewer than 3,000 were killed yearly in the 1920s compared with 7,000 in 1909. (The response was to increase the bounty from fourpence to sixpence per tail.) It's best to be cautious when inferring changes in wildlife population sizes from hunting returns, because what we might be measuring is the effort made by hunters rather than the abundance of animals. Nevertheless, the scale of the change is extreme, and it's confirmed by different sources.

People took a long time to start worrying about the conservation status of red squirrels. Leonard Harrison Matthews, writing what was then the definitive account in the New Naturalist series, *British Mammals*, in 1952, argued that reds were spreading in many areas and were 'not nearly so scarce as has frequently been stated'. Phew, that's all right then. As late as 1975, government scientist Andrew Tittensor was still advocating the control of red squirrels where they damaged plantations and reduced profit margins. At least he helped stop some of the persecution with his advice that control should be restricted to late spring and early summer, and only in areas of high-value timber. With numbers plummeting as the greys advanced, reds across Britain continued to be killed as pests until the Wildlife and Countryside Act of 1981 finally got round to giving them legal protection. In 1992, predictive models based on Forestry Commission surveys suggested a slight contraction in the red squirrel's range, but certainly no threat of extinction. Three years later, the Mammal Society sounded the alarm by announcing that reds were at risk of disappearing altogether south of a line stretching from Morecambe Bay to the Tees. That's when the panic started.

One consequence of this inglorious history is that the red squirrel subspecies peculiar to Britain and Ireland, *Sciurus vulgaris leucourus*, has been hunted and hybridised out of existence. In the nineteenth century, red squirrels were imported from Europe not only for reintroduction programmes but also as pets: 20,000 animals, including many derived from France, were sold in London during 1837 alone, where they became à la mode accessories for fashionable Victorian drawing rooms. One way or another, many ended up in the wild, where their genetic markers spread through the native population. Taxonomists can't even agree that *leucourus* ever existed: the verdict hinges on the significance (or otherwise) of a whitening of the tail in the summer months and a slight shortening of average skull height. This fussy debate has occasionally emboldened the most fanatical defenders of the greys to make the opportunistic claim that reds are themselves not really native. The rest of us understand perfectly well that a red squirrel is a red squirrel no matter what. We also know that, should circumstances ever allow, we can once again replenish British populations with red squirrels sourced from Europe.

*

Amidst these long-term declines, red squirrel numbers fluctuate wildly year by year. This is largely because they're heavily dependent on tree seeds for their nutrition, and species such as beech, oak, spruce and pine have strong cycles in their fruiting activity. When we lived at Corsham, a beech tree towered over our back garden. During what are called 'mast years' (from the German *mästen* – 'to fatten'), we couldn't walk on the lawn barefoot because the ground was covered in prickly shells. These bumper harvests are coordinated across vast distances, even entire nations, and they occur sporadically every five to ten years.

Masting remains one of nature's great mysteries. We don't fully understand why trees do it, especially given that it takes them a long time to recover from the massive energy expenditure and start growing normally again. The most popular theory is that masting is a result of co-evolution between trees and seed-eating animals. Normally, the abundance of squirrels, jays, boar and other species is kept in check by the modest amounts of food available. When a mast year arrives out of the blue, seed availability far outstrips consumption, ensuring that some of it survives to germinate. As an added bonus, squirrels use scatter-hoarding, burying seeds and cones in small quantities underground. They rely on smell to find their hoards, and what they fail to sniff out has been planted ready to grow.

Mast years trigger major upturns in red squirrel reproduction and survival. On average, not even a quarter of juveniles last long enough to celebrate their first birthday, but after that they have a 50–70% chance of survival year on year. In mast years, juvenile survival surges up to 50–60%, while adults become virtually bomb-proof.[3] Breeding in the spring following a mast year will start earlier, meaning that many adult females will produce two litters rather than one, and more of their offspring will survive to weaning. The strong impact of food availability explains why small, fragmented, species-poor woodlands common across the UK are so unwelcoming compared with the expanses of continuous forest found elsewhere in Europe. The fewer the trees, and the lower the diversity of species, the more likely that a population of squirrels will be vulnerable to food shortages.

We were battling through the half-term traffic; we'd timed our journey to Formby with impeccably poor judgement. As if things weren't bad enough already, this gave Tim ample opportunity to run through his extensive repertoire of George Formby songs, accompanied by air-ukulele. We turned up at 11 a.m. on one of those rare English days when the weather was too hot to be comfortable, and found the last spot in the National Trust car park. Red squirrels are most likely to be active a couple of hours after dawn, and again a couple of hours before dusk. They're not great lovers of midday sun, unlike the throngs of reddening beachgoers whose proximity must have made the squirrels even more inclined to keep a low profile. Despite having needlessly lengthened our odds of success, we remained optimistic for no better reason than that several friends had recently reported watching reds here.

This patch of coastline was once a saltmarsh and at low tide it's possible to see, where erosion has exposed the clay, 7,000-year-old prints of aurochs, red and roe deer, oystercatchers, wild boar, dogs and humans. The woodland at Formby is nothing like as ancient. It consists primarily of non-native conifer – no surprise there – and much of it was planted towards the end of the nineteenth century to fix the sand dunes or at least slow their erosion. This is where the reds live, and for the avoidance of any doubt a popular trail is clearly signposted as the 'Red Squirrel Walk'. A National Trust volunteer greeted us and explained that there was no particular squirrel hotspot: they were randomly scattered throughout the woodland. They used to congregate in the same area, because photographers regularly placed peanuts on tree stumps near the car park to entice them down from the canopies. This is now discouraged because it doesn't do much for their diet, their social dynamics or their physiology: weakened by the easy pickings, Formby's red squirrels have a less efficient temporal muscle for chewing than any other assessed population in Britain and Europe.

Within fifty metres, we came across a plywood signpost on which the Lancashire Wildlife Trust had stapled a message ominously titled 'Squirrel Pox Warning'. The small print wasn't encouraging, either: 'We are receiving regular calls about sick and dead red squirrels with pox symptoms in Formby. This has now spread from Lifeboat Road to Victoria Road.' The virus was killing reds at both ends of the woodland, and presumably everywhere in between. People were urged to report sightings of diseased squirrels; to assist that grim responsibility, the message ended with photographs of infected individuals and a caption that drew attention to 'swollen and/or crusted eyes'. That's probably the most conspicuous symptom of squirrelpox, but there are others no less appalling. The virus causes lesions around the eyes and mouth, blinding the squirrel and making it unable to eat. The animal will shake and struggle to breathe, its fur will drop out, and there's likely to be discharge from the nose and eyes. Squirrelpox usually kills within four or five days, although in the cruellest cases the decline can take up to two weeks. Squirrels must have been suffering horrific deaths at that moment, secretly and silently, out of sight high up in the trees as we wandered past cheerfully below.

The Covid pandemic demonstrated just how hard it can be to understand the mechanisms by which virus spreads. In the midst of a global emergency, scientists failed to agree on the efficacy of facemasks, the likelihood of catching disease by touching surfaces, the relative risks of indoor and outdoor spaces, the usefulness or otherwise of lockdowns, and the safe social distance at which we should navigate around each other. Similar uncertainties apply in studying the transmission of squirrelpox virus. We can't require squirrels to wear facemasks or follow the two-metre rule, but if we work out exactly how the virus is passed between greys and reds, and between reds and other reds, there may be strategies for intervention.

One recommendation was highlighted in yellow at the bottom of the Lancashire Wildlife Trust's sombre notification: 'Please remove all feeders completely as pox virus is readily transmitted between squirrels using feeders.' Not only will greys and reds converge on these tasty treats and possibly come into contact with each other, but a newly infected and still-active red could easily contaminate the entire population by shedding the virus along the busiest squirrel highways and meeting-points. Squirrelpox is a textbook case of disease-mediated invasion: a virus spills over from the non-native host species into its more susceptible native competitor, speeding up the process by which one ousts the other. It used to be believed that greys were driving reds out by beating them up in physical confrontations. We now know that the two species tend to ignore each other, and that although greys are much bigger (520 versus 300 grams), on those rare occasions when things do turn nasty the red is just as likely to win.

Scientists have identified two possible routes for squirrelpox transmission. Reds and greys share fleas, such as the species *Orchopeas howardi* that was introduced into the country by grey squirrels. Fleas have a long and ignoble history of transmitting pathogenic diseases, from Black Death to myxomatosis, and the seasonal fluctuations in squirrelpox virus mirror the patterns of flea infestations. Recent research has discovered an even more important pathway – the squirrel's forearm.[4] Squirrels have a patch of sensory hairs and glands on their arms that are crucial for scent-marking behaviour, and swabs taken from that area carry viral loads about 250 times higher than the blood, mouth or anus. Viral shedding from the arm gland can go on for months at a time. A garden feeder turns out to be the perfect place to pick up something deadly that's been left behind by a previous visitor.

Estimates of the proportion of grey squirrels carrying the virus vary from four percent to as high as seventy percent. This may capture genuine fluctuations from one part of the country to another. It took decades to establish that squirrelpox virus was the main driver for the red squirrels' decline, and one reason for the delay was the existence of curious anomalies. There'd been a previous population crash among reds in the early twentieth century, which couldn't have had anything to do with grey squirrels because they weren't present across much of the country at that time. John Gurnell, author of what may still be the finest book written on squirrels, summarised the consensus when he concluded in 1987 that 'there is no evidence that grey squirrels have brought with them a disease which is causing the downfall of the red'. He was looking at examples like Thetford Forest, where greys and reds had reportedly lived side by side for twenty years. Maybe those greys, by chance or some quirk of their environment, carried a lighter viral load. The happy equilibrium didn't last. Despite a massive culling campaign against the greys, the reds had completely disappeared from Thetford Forest before the end of the century.

The scientific argument for the crucial role of grey-to-red transmission of squirrelpox virus is now proven (partly thanks to Gurnell's own subsequent research). Anti-cull campaigners will sometimes refer to cases where outbreaks have taken place without greys being present locally, but those accounts tend to be anecdotal and, even if accurate, don't refute the basic premise. We know that greys can hide under a car bonnet, so why shouldn't a single male turn up in a new area and infect the local population of reds before dying a lonely bachelor? Left to their own devices, an established colony of greys will spread at the rate of roughly eighteen square kilometres each year, but the reality on the ground is far more haphazard, and DNA testing uncovers all sorts of mysterious geographical leaps. For example, the grey squirrels of Aberdeenshire are much more closely related to the greys of the New Forest than to any colonies in between. Somehow, and at some indeterminable point, greys have made a 570-mile journey from one end of Britain to the other, taking squirrelpox virus along for the ride.

Tim grew up in Plymouth, which hadn't seen a red squirrel since before the Second World War. Fiona had better luck in North Lancashire. As a child during the 1970s, she'd often visit Squire Anderton's Wood on the outskirts of Preston to pick bluebells and spot squirrels. Those squirrels were red; nobody had ever come across a grey one. Today, Squire Anderton's Wood lies to the north of the ever-widening M6 motorway – one of the most congested sections of road in the country – and it still has bluebells, but the red squirrels are long gone. Ask a local for directions to 'Squires Wood' and you'll be sent to a new road running through a nearby housing estate (where it joins other roads whose names rub our noses in what's lost: 'The Briars', 'The Brambles', 'Ivy Bank'). To the south of the motorway, another of Fiona's childhood haunts, Clough Copse, is getting strangled by its surroundings, although a local newspaper describes the surprising tranquillity to be found on a woodland walk enclosed by a 'depotscape' of industrial units and housing. Further along, what was once an ancient semi-natural woodland has been squeezed into a 100-metre strip to make room for a golf course and more housing. 'They paved paradise,' Joni Mitchell used to sing, 'and put up a parking lot'; maybe she had Preston in mind.

Could we reintroduce reds into places like this? Not really, because the red squirrel is a creature of tree canopies. It spends at least seventy percent of its time off the ground – a far higher proportion than the greys – so habitat fragmentation poses an insurmountable problem. Research in Lancashire during the 1990s showed that woodlands were unlikely to contain red squirrels if they were more than five kilometres from a major red squirrel colony;[5] reds will sit tight rather than be tempted to pioneer their way across deforested landscapes. Grey squirrels, on the other hand, spend a lot of time on the ground, particularly in the winter, and are evidently content to scurry between the scattered trees of parks and gardens and along hedgerows. Even if we persuaded the locals to support the culling of greys for miles around, they'd be back.

It's convenient to blame the greys for everything and ignore our own destruction of habitats. But forget the greys for a moment: where would red squirrels find a reliable food supply amidst this horribly fragmented landscape? In years of poor nut harvests, even if starvation didn't finish them off, their weakened immune system would leave them vulnerable to disease. We could perhaps develop a programme of supplementary feeding, but does a population trapped in a ghetto, entirely reliant on handouts provided by humans, deserve to be described as wild? We might just as well round them up and deposit them in the grounds of a stately home, where we can observe them over a polite cream tea. Which, funnily enough . . .

*

Down the road from us at Escot Park, there's a captive breeding facility and visitor centre, developed with Heritage Lottery funding. The public is able to wander through a boarded walkway and enjoy watching the red squirrels. By now, those squirrels aren't at all perturbed by humans. We used to visit regularly when our children were small, and on one occasion a squirrel launched itself onto a friend's arm, ran up her back and landed on her son. Staff blamed this impertinent behaviour on our yellow raincoats, which looked similar to the coats worn by keepers at feeding time. One ice cream later the child had forgotten his trauma, and we continued on our merry way to visit the wild boar.

Many red squirrel conservation initiatives have involved the reintroduction or translocation of squirrels. Captive breeding can sometimes help with reintroduction programmes, but in the case of red squirrels it's fraught. Recently, captive-bred squirrels have been released in the north of Scotland, sometimes in areas beyond their natural range. This range expansion has then been presented as a success story, despite breaching the IUCN's guidelines on reintroductions. The genetic health of captive-bred populations is also rarely considered. Squirrels are oblivious to taboos against incest and, in no time at all, large populations of animals can be descended from a tiny number of founders. Why reintroductions prefer naive captive-bred animals over wild stock sourced from the plentiful supplies in continental Europe is a mystery.

As a minimum, there needs to be a central body with strategic oversight of captive breeding. Instead, a variety of organisations with different standards and ambitions are doing their own thing for all sorts of species, squirrels included. As long as they aren't taking animals from the wild (which would need a licence), anyone can breed red squirrels. One obvious drawback is that keeping wild animals on public display where they'll become habituated to people prevents any prospect of them developing the skills needed for survival on the other side of the enclosure. Losing a fear of humans and other predators is never a great strategy. To its credit, Escot ensures that there are additional areas away from public view, but not everyone is so considerate. Red squirrels in captivity are also prone to developing stereotypies – repeated behaviours such as running back and forth along the cage. Those animals wouldn't last five minutes in the wild.

An even bigger concern is that breeding centres keep animals at unnaturally high densities, and disease transmission can take place between squirrels and with other captive species. Red squirrels are vulnerable to all sorts of deadly pathogens. The parasitic gut infection coccidiosis, still turning up in post-mortem examinations of both red and grey squirrels, has been responsible for mass die-offs in the past: a million reds in Finland in 1943–4, and huge unspecified numbers of grey squirrels in England in 1931. Even worse is adenovirus, which causes severe intestinal disease. It's been found in twenty-three out of the twenty-six tested breeding centres, having probably been spread by the exchange of animals.[6] The virus is now present in both red and grey wild squirrels in Scotland, and the suspicion is that it arrived via reintroduction programmes. The headline act for squirrel diseases, though, is leprosy – exactly the same strain, in the Brownsea Island population, that killed humans in the Middle Ages.

*

Some facts to squirrel away before the next pub quiz.

The Latin *sciurus* derives from the Greek *skia* (shadow) and *oura* (tail). 'Shadow-tail' doesn't immediately make a lot of sense, but the name probably refers to the tail's tendency, when a squirrel sits upright, to act as a sunshade by curling over the animal's back. One dialect word for squirrel is 'skug': 'shelter'.

The Old English name for a squirrel was *acweorna*, which means 'oak-defender' or 'oak-shelterer'. The English word 'acorn' looks like it must be hiding amidst this etymology, and seems even more obviously present in some of the modern Germanic words for squirrel, such as *eekhoorn* in Dutch or *egern* in Danish. But apparently not: in linguistic terms at least, squirrels have nothing to do with acorns.

There are close to 300 species of squirrel across the world, and the United States alone has 65 separate species. It's a testament to the evolutionary success of the basic template. Squirrels can be roughly subdivided into six groups: tree squirrels, ground-dwelling squirrels, flying squirrels, marmots, chipmunks and prairie dogs. Flying squirrels can't *really* fly. Like Buzz Lightyear, they're falling with style.

'You can drop a mouse down a thousand-yard mine shaft,' says the scientist J.B.S. Haldane, 'and, on arriving at the bottom, it gets a slight shock and walks away, provided that the ground is fairly soft. A rat is killed, a man is broken, a horse splashes.' It would be touch-and-go for a tree squirrel to survive an impact at terminal velocity, but many of us have seen them tumble out of high canopies onto tarmac and scamper off, barely discombobulated.

In the fourteenth century, Edward III ordered a counterpane made out of 2,240 squirrel skins so that he could get a comfortable night's sleep. Squirrel served the luxury end of the fur market. Records show that thousands of red squirrel skins were exported from Ireland in the sixteenth century, the trade stopping only when they ran out of squirrels.

When they weren't being turned into bedspreads, red squirrels were kept as pets by highborn ladies. They often appear as stylish accessories in medieval paintings, and are themselves accessorised with silver chains and collars. There seems to have been a bawdy French tradition in which squirrels became part of an elaborate code for sex. The lady's virginity was a nut that needed to be cracked, and what more fitting animal to symbolise that task than a squirrel?

Fashionable Victorians and Edwardians used red squirrel tails to decorate hats, and lined collars and cuffs with squirrel fur. Well into the 1970s, there was still a niche

UK market for squirrel fur paintbrushes and angling flies, although it's doubtful that this trade had any impact on the numbers hunted. In 2005 alone, 170,000 individuals were trapped and killed in Mongolia, many of which ended up in traditional Chinese medicines.

Fiona reports that when you handle an anaesthetised squirrel, the first thing you notice is that it seems to be all muscle. Its body is hard and there's no give anywhere. By comparison, a rabbit – even a wild one – feels saggy like a bag of custard.

Red squirrels have two moults. The spring one starts from the head, gradually working its way towards the rump. The animals look scruffy and can develop bald patches until their summer coat grows through. The autumn moult goes in the reverse direction. The tail and ear tufts only moult once a year.

The fact that a recent Mammal Society meeting involved a 'Spot the red squirrel' competition – the other exhibits being of grey squirrels, not lions or walruses – proves that even experts can struggle to distinguish reds from greys. Red squirrels undergo remarkable colour changes as their bright orange-red fur of early summer gives way to dull brown or even greyish colours later in the season, and greys sometimes go in the other direction by growing extensive red patches on their summer coats. Greys don't have ear tufts, but neither do reds in late summer. Despite the hazardous ice-cream vans that often seemed to feature, those road safety campaigns starring Tufty the impossibly large squirrel must have been set in winter when ear tufts are particularly prominent.

Squirrels can be left- or right-handed. They gnaw a pine cone from the bottom up, twisting it as they go. This means that you can distinguish handedness from the pattern of the spirals. For left-handed squirrels, the spirals will turn anticlockwise, and clockwise for the right-handers. Researchers believe that problem-solving skills are greater in more ambidextrous squirrels than in those with a strong preference for the left or right paw.

'What is a squirrel but an airy pig?', John Keats wrote to a friend in February 1818. Keats was an opium addict.

'This is a Tale about a tail – a tail that belonged to a little red squirrel, and his name was Nutkin.' The perfectly measured rhythms of Beatrix Potter's opening line introduce one of her finest stories. Nutkin is a naughty squirrel. While his brother Twinkleberry and his many cousins bring offerings for Old Brown the tawny owl, and deferentially seek permission to collect acorns and hazelnuts from Old Brown's island, Nutkin is rudely intent on winding the owl up. He bobs like a cherry, tickles Old Brown with a nettle, or dances up and down like a sunbeam, while doing his best to antagonise Old Brown with silly riddles. We all had a Nutkin in our class at school, but here he remains recognisably a squirrel with squirrel-like behaviour. (Potter herself had bought two squirrels for inspiration, only to get rid of one because it kept fighting the other and ripped part of its ear off.) Nutkin finally pushes his luck too far. Old Brown flips, pins him down, and is about to start skinning the squirrel when Nutkin manages to escape his clutches, leaving half his tail behind. 'And to this day, if you meet Nutkin up a tree and ask him a riddle, he will throw sticks at you, and stamp his feet and scold, and shout – "Cuck-cuck-cuck-cur-r-r-cuck-k-k!"'

Nutkin and his family members reach Old Brown's island by making little rafts out of twigs. They each have an oar, and they use their tails as sails. Weirdly, there's quite a long tradition of sailor-squirrels in literature. One seventeenth-century source describes a squirrel seeking out 'some rinde or smal barke of a Tree, which she setteth upon the water' before navigating with her tail. A century later, Oliver Goldsmith describes Lapland squirrels taking to the water, 'every squirrel sitting on its own piece of bark, and fanning the air with its tail, to drive the vessel to its desired port'. When the weather gets rough, the squirrels drown and are skinned and eaten by Laplanders as their bodies wash ashore.

One other literary squirrel deserves a mention. In Norse mythology, Ratatoskr runs up and down the world tree, Yggdrasil, conveying messages between the eagle at the top and the dragon at the bottom. He seems to be a proto-Nutkin, because he enlivens his journeys by telling slanderous gossip at both ends. Red squirrels, we suspect, are annoying just for the hell of it.

*

What's *most* annoying about red squirrels is their destruction of trees. As well as nibbling at tree-shoots, they chew the bark, causing calluses and leaving low-quality scarred timber. If the squirrel chews all the way around a trunk – known as girdling or ring-barking – it's curtains for the tree because the arterial flows get severed. Greys are even bigger vandals and, according to estimates produced for the Royal Forestry Society, they're likely to cost forestry in Britain £27.5 million year on year (a total of £1.1 billion) over the next forty years.

The greys' antisocial behaviour seems to have been picked up on this side of the Atlantic. Back in their native habitat, they only chew the bark of a few tree species, such as maple with its irresistible sap. One explanation for their different habits here is that squirrels in Britain have a nutritional deficiency in calcium not found in North America, and bark is a particularly good source at the time when it's eaten by squirrels.[7] Thin-barked trees are targeted most often, and damage is worst between April and early July in places where populations are high. Grey squirrels often single out the most vigorous and dominant trees in a stand; as if that weren't annoying enough for foresters, money spent on squirrel control doesn't always produce a corresponding decrease in damage.

Michael Heseltine caused a stir a few years ago when he reported having killed 400 grey squirrels on his seventy-acre Northamptonshire estate over a nine-month period. Despite the inevitable backlash, Heseltine's only crime was to describe bluntly what happens on a routine basis up and down the country. Even organisations like the RSPB and the wildlife trusts kill grey squirrels, although they're publicity-conscious enough to prefer descriptions like 'managing', 'controlling' and (slightly riskier) 'removing', and the wildlife trusts tend to cull only while protecting populations of reds. Setting aside the emotive ethical debates, this effort comes at huge expense: Heseltine didn't really do any killing himself, but employed two gamekeepers. That's an awful lot of money per dead squirrel, and it doesn't stop there. You have to cull every year, all year round, ad infinitum if not ad nauseam, because otherwise the greys will recolonise from surrounding areas. By all means rail Cnut-like against the incoming tide, but it's doubtful that you can kill them fast enough or for long enough.

The strategy that Andrew Tittensor proposed for red squirrels – to control them at the right time of year and focus on locations with the most valuable timber – also works well for greys wherever the objective is solely to protect timber crops. Exterminating grey squirrels altogether, or even keeping their populations depressed, requires investment of a different magnitude. No plan survives contact with the enemy, especially when the enemy's a squirrel. Simple rules of population dynamics mean that it's perfectly possible to kill thousands without having any long-term impact: the remaining individuals survive and breed better because there's less competition. A ready supply of grey squirrels awaits in parks, back gardens and hedgerow trees, and they're all too willing to bound into the empty territories created by culling programmes. Drop your guard momentarily and you've got as many squirrels as you started with.

Standard approaches have mostly failed to halt the invasion. A survey of twenty-seven red squirrel conservation projects reached the unsurprising conclusion that it was hard to recruit volunteers and maintain their commitment for the cage-trapping of grey squirrels. Rain or shine, you have to go out and check traps twice a day to kill any captured animals. Most nature lovers prefer to monitor reds than to kill greys, and we might start to doubt the motives of anyone who showed too much enthusiasm for the bloodshed. As the founder of one group admits, 'We only call ourselves the Red Squirrel Protection Partnership because if we called it the Grey Squirrel Annihilation League people might be a bit less sympathetic. But we do nothing with red squirrels apart from save them by killing grey squirrels!' Even the people who do the killing tend to dislike the tedium of cage-trapping and the unsporting close-range dispatch, but free-shooting of wild squirrels is nowhere near as effective.

Many projects lurch from one short-term source of funding to another. It's almost impossible to get funding for ongoing routine work, however necessary. The 'novelty-driven' strategy of the Natural Environment Research Council (NERC) means that there's never a snowball's chance of funding critical monitoring projects. When it comes to culling grey squirrels, funders are also scared away by worries about adverse publicity. The final and most perverse problem of all is that it can become increasingly difficult to motivate volunteers as soon as success approaches. It's dull and thankless to trail round a woodland when there are hardly any grey squirrels left, but that's precisely when vigilance to avoid new incursions is vital.

The difficulty of establishing an effective cull strategy is illustrated by the history of Red Squirrels United, a conglomeration of various wildlife trusts working with Newcastle University and the research arm of the Forestry Commission. Their £3 million project, funded by the EU and the National Lottery, began with the aim of protecting and bolstering red squirrel populations in nine sites: four in Northern Ireland, three in Wales and two in England. During 2015, 21,000 grey squirrels were culled in northern England, at a cost of £60 per squirrel, or £1.26 million in total. If we were given £3 million to spend on biodiversity, it's safe to say that killing squirrels wouldn't feature high on our list of priorities. Their budget was vastly larger than the Mammal Society's, the British Dragonfly Society's, and many of the wildlife trusts'. Red Squirrels United came to an end in 2020 when the money stopped, but the greys carried on going: they're breeding at a rate in excess of what even a well-organised and amply funded project like this had any hope of controlling.

Misery loves company, and for once Britain's self-inflicted wildlife wound is matched by one of our continental neighbours. Italy is the only other country in Europe to have grey squirrels. In fact, the Italians lay claim to not just one invasive squirrel, but four. First introduced to Piedmont in 1948, and to Liguria in 1966, the grey squirrel was joined in the 1980s by Finlayson's squirrel from South-East Asia. Another South-East Asian interloper, Pallas's squirrel, arrived in the past decade, along with several populations of Siberian chipmunk. At the very least, Britain should provide an object lesson in what other countries must never do, but the popularity of squirrels as pets and ornamental additions to public parks seems to continue unabated in Italy. As late as 1994, the town of Trecate funded the release of three pairs of grey squirrels – yes, it paid good money to create an ecological disaster all of its own – only to recapture them two years later after pressure from exasperated scientists.[8] The good news for the Italians is that, by some freakish luck, their grey squirrels don't seem to have brought squirrelpox with them.[9] The rest of the news is pretty dismal.

The Italians may not have learned from us, but maybe we can learn from them. Their experience provides the answer to a useful thought experiment: what would happen if we could eradicate squirrelpox? We now know that it would slow but not solve the problem. Greys oust reds at a rate that's between seventeen and twenty-five times faster when they're carrying squirrelpox, but even when they're not, they'll win in the end.[10] Greys strip hazel trees bare before the fruit is ripe – and before reds get round to collecting nuts and caching them for the winter. Reds don't fare any better when it comes to acorns, which they can't digest as effectively as can greys. In all cases where the two species are competing for resources, the reds come off worse; their smaller size and lower autumn weight also exacerbate their vulnerability to energy shortages.[11] It's true, though, that reds cope less badly than usual in coniferous woodland, particularly spruce, because greys aren't so well adapted to feed on cones. Coniferous woodland is being used as a critical barrier that can be defended by culling programmes, but even here juvenile red squirrel survival is reduced to thirteen percent where greys are present.[12] The *coup de grâce* is that greys pilfer the food stashed away by red squirrels.[13] They pilfer each other's hoards, too; greys are such rascally varmints that they spy on each other in nut-burying season, and if they know they're being watched they'll dig holes and only pretend that they're leaving food in them. Not even Nutkin's as cheeky as that.

For twenty-odd years after their release in Piedmont, the grey squirrels hung around within an area of about twenty-five square kilometres, and avoided getting themselves into too much mischief. Then as soon as the population became established and competition for territories ramped up, colonisers set out across the inhospitable agricultural areas that surrounded the release zone. The grey squirrels' range started doubling every five years or so, and between 1970 and 1990 it increased more than tenfold. It was only a matter of time before the squirrels reached the sunlit uplands of eastern Piedmont's alpine broadleaved forests. By 1999 they were expanding across 250 square kilometres a year – a rate higher than anything seen among established populations on our side of the Channel.

With our own litany of disasters, we Brits aren't in any position to roll our eyes. Nevertheless, we can fairly wonder how this could have been allowed to happen. As we know from the Covid pandemic, the checking of exponential growth rates requires early action. Wait too long and numbers become unmanageably vast. So let's travel back in that time machine to

1996, when there were between 2,500 and 6,400 greys in Piedmont. Their population was still enclosed by an agricultural landscape that offered only small and fragmented woodlands. The edge of their range was seven kilometres away from continuous alpine woodlands, and modelling predicted that they would arrive there in a year or two.[14] Emergency action was needed straight away, because otherwise the grey squirrel would be bounding unstoppably across northern Italy into France, Switzerland and the rest of continental Europe.

Mindful of the attempts to halt grey squirrel invasions in Britain, the Italian Wildlife Institute had started a trial eradication project that focused on a small grey squirrel population in Turin.[15] The objective was to work out which control strategy would be most effective and efficient, all the while keeping an eye on the spread of grey squirrels elsewhere, and planning an eradication programme that would work across the entire region. Early results suggested that control was indeed achievable. Everything was going to be all right – until, that is, court action initiated by an animal rights organisation brought the programme to a screeching halt. The Italian legal system isn't renowned for its speed and, although the Italian Wildlife Institute would eventually be exonerated, the court proceedings and judicial enquiry took three years. By that time, grey squirrels had spread across such a massive area that an eradication programme was no longer possible. All efforts have now switched to limiting the expansion in areas considered most critical for the conservation of red squirrels. The best-case scenario – already proven by their recent progress to be hopelessly optimistic – is that grey squirrels won't reach the western

Alps until 2026 to 2036, Switzerland until 2051 to 2066, and France until 2066 to 2071.[16]

Research indicates that city-dwellers' attitudes towards alien species are generally positive, whether those species be rhododendrons, Canada geese, or the flocks of rose-ringed parakeets now squawking their way over London and many other European cities. There's an added complication where native species are coexisting with the non-natives. Scientists in Genoa grappled with exactly this problem in 2015. How could they protect native red squirrels in urban parks where the locals had started feeding greys? Cage-trapping followed by shooting didn't seem like a viable option in a public space, while lethal traps or poison risked accidentally bumping off the reds that they were trying to save. The obvious answer was birth control.[17] A survey in Britain has already suggested that the public find this option more palatable than culling: 61.5% call it 'acceptable' or 'very acceptable'.[18] There could even be side benefits. Because a neutered squirrel can live for between five and seven years in child-free bliss and maintain a territory successfully during that time, birth control doesn't bring about the sudden influx of immigrants from surrounding areas that culling regimes usually provoke.

Given the need to act quickly, the Italians went for a direct approach and physically neutered the animals. Teams working in Genoa's urban parks caught 321 grey squirrels, male and female, which a vet then sterilised. (We've tried and failed to come up with a joke about taking away their nuts.) As expected when dealing with wild animals, some didn't survive: three squirrels died before surgery, and ten post-operatively. After a couple of days of convalescence, the rest were released into new parks, and monitoring showed that 94.5% of the ones that avoided getting squashed on the roads were still alive two months later. What happened after that? Did the neutering squads round up grey squirrels across the urban parks of Italy? Were the red squirrels saved? Not exactly. The animal rights backlash became too hot for the government, and all funding was suddenly withdrawn.

Here in Britain, government agencies have occasionally raised the possibility of squirrel family planning, and some overexcited journalism has suggested that we're on the verge of fixing the grey squirrel problem once and for all. Sadly, we aren't. There are currently trials underway for two different oral contraceptives. The first is a vaccine that suppresses reproductive hormones. In an injected form (GonaCon™), this has been successfully used in animals including feral goats, donkeys and horses. The second is a drug that works by inhibiting cholesterol (DiazaCon™), with some promising results when fed to squirrels.

Nobody wants to have to catch the nation's squirrels and inject them, so efforts are focusing on feeding oral contraceptives. One of the major problems is ensuring that animals eat sufficient bait for it to be effective; squirrels are territorial, so the job of exposing enough of them to the bait is also fiendishly complex. Trials in laboratory rats of an oral vaccine similar to GonaCon found that sixty percent infertility could be achieved, but the animals needed six doses within a month.[19] Different formulations were even more demanding, in one case requiring daily consumption for ten days. How, or whether, squirrels can be persuaded to feed on bait sufficiently to become infertile remains an open question. The amount of bait taken no doubt varies with time of year and the availability of other resources. Food is the music of love, and in years when oak or beech mast is bountiful – exactly when there's plenty of squirrel sex going on – the incentive to take bait from feeders is much lower.

The next challenge is to avoid unwanted effects on other animals. Birds operate a different hormonal system, so they won't be at risk. It's also unlikely that there'll be any problem for predators such as pine martens, because digesting the squirrel will render the hormones inactive. The same can't be said for dormice, wood mice, yellow-necked mice, bank voles, and those red squirrels on whose behalf we're making the effort. All are native, and all visit peanut feeders and other food dispensers; how do we exclude them from a bait delivery system that's large enough to admit a grey squirrel? Researchers are currently trying to design a device that will only admit animals of the right weight.

Defra suggests that when used on low-density populations after culling programmes, these fertility inhibitors could achieve roughly the same results as lethal control.[20] It hardly sounds like they're anticipating the revolutionary impact that media stories might have you believe. Contraception may well be helpful in parts of the country where greys haven't yet established themselves, or at the fringes of the distribution where populations have been reduced already by culling, but it's going to be at least 2026 before squirrel contraceptives are available for widespread use.

Like culling, squirrel family planning only works if there's a long-term programme. Bear with us while we prove this point by talking about . . . feral Kashmiri goats on the Great Orme headland in North Wales. Working with Defra, a team of Marine commandos rounded up as many of the animals as possible for injection with GonaCon. (One of them complained that it was a tougher assignment than Afghanistan.) This was the first example in Europe of real success with a free-living mammal population: fertility was reduced substantially in year one, depressed to a smaller extent in year two . . . and back to normal in year three. The project had come unstuck when Covid stopped the annual round-up, producing a series of headlines that sound like a Chris Morris fever dream: 'Llandudno goats cause chaos after skipping contraception' (ITV News); 'Rogue herd of goats take over town after contraception stopped' (*Daily Express*); 'North Wales town taken over by large herd of goats' (Fox News).

There's no population explosion among red squirrels in Formby. They're rare enough to be extremely difficult to spot. The best policy is to walk quietly and then pause for five or ten minutes. Picnic breaks work well; Fiona's closest encounters as a child came when she was sitting and eating a sandwich. You don't spot a squirrel by moving purposefully. Just keep still and, if you're lucky, it'll reveal its whereabouts by chomping noisily on a pine cone, getting into a territorial dispute with a neighbour, or scuttering around at high speed. Embarrassing though it is to admit, we knew all of this when we arrived, and we realised immediately that we'd bungled it. There were too many people in the reserve. The walk was pleasant enough, but we weren't going to see any squirrels. Overheated and with the money running out on the parking meter, we grumpily admitted defeat and trudged back to the car, vowing to return another time. Let's hope that Formby's red squirrels survive long enough to be still there waiting for us.

EIGHT

A Phocine Good Story: Saving Grey Seals

African elephants are rare, right? About 350,000 savanna elephants live wild across the continent, numbers having declined by sixty percent over the past half-century. Elephants have become the blockbuster act for money-raising crusades to save species threatened with extinction. After all, nobody brings in the big bucks by fronting their campaign with the Hispaniolan solenodon. But what if there were a species in Britain as rare as the savanna elephant? What if this species were highly intelligent, with complex social interactions and a life expectancy of twenty-five years or more? And what if thirty-six percent of its global breeding population – and eighty percent of its European population – could be found here on our shores, from the Isles of Scilly to Shetland? In those circumstances, no doubt we'd do everything in our power to conserve that animal, grateful for the opportunity to demonstrate the richness of our native wildlife to an appreciative global audience. Right again? We'd never dream of granting licences to shoot pregnant females or mothers with dependent offspring. Apart from anything else, how could we condone that butchery while hectoring poorer nations about the importance of wildlife conservation?

You'll have guessed already where this is heading. The species in question is the grey seal, and there's a simple reason why you won't find seals in the 2018 population review, despite Fiona's request to include them. They give birth on land and spend about a third of their time lolling around on land, but they're treated by government agencies as a marine species alongside the cetaceans (whales, dolphins and porpoises). We think that's a bit weird: after all, a coastal otter hunts in the sea but it's definitely a terrestrial animal, and seals seem far closer to otters behaviourally than they are to whales. Under other circumstances, it wouldn't matter which label you gave them, but at the moment there's no British Red List for marine mammals.

We're an island nation with legal obligations to fulfil. As signatories to the international OSPAR Convention, the British government must ensure that seal numbers don't fall by ten percent over any five-year period or by twenty-five percent in the longer term. Monitoring mammals at sea is harder than on land, but seals spend so much time ashore that counting them is relatively straightforward compared with dolphins or whales. As it happens, we already have detailed estimates of our seal populations, thanks to the regular surveys carried out by the Sea Mammal Research Unit at the University of St Andrews (SMRU to its friends, who somehow pronounce it as a single word). Their work is overseen by a Special Committee on Seals (SCOS), which advises the government. So it wouldn't take much effort to add our two native seal species to the population review. Better still, the government should commission a new official census for all our marine mammals.

Seals are pinnipeds – from the Latin meaning 'fin-footed'. In fact, the word *pinna* can refer generally to any projecting body part, such as an ear-flap, a feather or a wing, so sometimes you'll see 'pinniped' glossed as 'wing-footed' or 'flap-footed'. In common with most marine mammals, seals' ancestors were land-based carnivores. As Charles Darwin speculated in *The Origin of Species* long before any fossil evidence proved him right, 'A strictly terrestrial animal, by occasionally hunting for food in shallow water, then in streams or lakes, might at last be converted into an animal so thoroughly aquatic as to brave the open ocean.' Whereas the cetaceans pursued this skill to its logical conclusion and gave up terra firma altogether, the amphibious pinnipeds are at home in either element.

There are thirty-four species of pinniped across the world, and these can be broken down into three categories: the *Odobenidae*, of which the walrus is the sole survivor; the *Otariidae* or eared seals, which include sea lions; and the *Phocidae* or earless seals, sometimes known as 'true seals'. Our two native British species – the grey and the harbour seal – both come from this last group, hence the adjective 'phocine', meaning 'seal-like'. When Fiona and her international colleagues were setting up Mammal Conservation Europe, they had to reject the designers' first draft logo because it depicted seals with visible ears; if you spot one of those around these parts, it's in the wrong ocean and seriously lost.

The walrus and the otariids are front-wheel drive, propelling themselves through the water with their fore flippers; phocids scull with their hind flippers. This has consequences for land speed. Over short distances, sea lions are capable of bounding along faster than Usain Bolt (fifty-six kilometres per hour against a mere forty-five for slowcoach Bolt), and they're actually faster on land than in water. Phocids are much slower because they can't invert their hind flippers under their body. They discover very quickly that a sculling motion doesn't work out of water, so they're reduced to doing the worm – they wriggle rather than walk. Having only ever encountered phocids, Aristotle called the seal 'a kind of imperfect or crippled quadruped', judging it by its awkward performance on land.

Of our native British species, the harbour seal is *Phoca vitulina*, meaning 'plump calf', while the grey seal is *Halichoerus grypus* – 'hook-nosed seapig'. These expose a failure to see wild animals as anything other than wonky deviations from domestic livestock. If you hear someone referring to a common seal, they're talking about the harbour seal, which is commonly called the common seal despite being less common here than the grey seal, which is less common globally. Got that? SMRU's recent estimates for British populations give roughly 157,000 grey seals and 44,000 harbour seals living around our coastlines. These numbers aren't directly comparable. Grey seal pups can be easily distinguished and excluded from surveys, whereas harbour seals don't congregate for pupping: their pups are capable of swimming straight away and are counted along with the adults later in their development when they come ashore to moult. What's most impressive about the seal census is the tiny margin of error: grey seal numbers are give or take 20,000, and harbour seals give or take 6,000, which may seem like a lot but is remarkably precise compared with population estimates for just about any other British mammal. But then, most other species don't helpfully lie about while you count them (or, easier still, take photos of them by light aircraft or drone).

Spot a seal in Wales or on the south or west coast of England and it's likely to be a grey. In Scotland and on the east coast of England it's more of a coin-toss. Harbour seals are smaller than grey seals – less than two metres long and about half the weight – with a short muzzle and a prominent forehead. A grey seal looks like someone has grabbed hold of its nose and given it a good hard tug. This extends the muzzle and slopes the forehead, making the grey seal appear rather canine; our chocolate Labrador, Charlie Brown, looks much closer to a grey than a harbour seal. People talk about differences in pelage (the seal's fur), but we've seen so much variety in both species that, for identification purposes, we'd hesitate to trust colour or patterning. Never assume that grey seals are grey, or that harbour seals aren't.

The grey seal is the largest wild animal to breed on British soil (or sand or rock). A bull seal might weigh 240 kilograms and a cow 160 kilograms; in Canada they're even bigger, with a dominant bull reaching 300 kilograms or more. That's about the same weight as a grizzly bear. As these numbers suggest, it's simple to tell male and female grey seals apart when they're lying next to each other, because with males being fifty percent heavier, sexual dimorphism is particularly pronounced. (This is nothing compared to the southern elephant seal, where the bulls can sometimes weigh ten times more than the cows.) The biggest males drive off their smaller rivals and take up position amidst a group of females. Those females are invested in mating with alpha bulls, because larger offspring are likely to be more successful. Even when they're physiologically receptive, they make a huge commotion about mating, probably because the noise attracts other males and ensures that only the most dominant will mate with them. For the bulls, this is a high-stakes game that can lead to serious war wounds; their health isn't improved by the need to stay ashore and defend their patch for the entirety of the breeding season, which means no food for at least eight weeks. Unsurprisingly, many of the smaller bulls opt out altogether, and lurk in the shallows for an opportunity to mate when the females take to the water. Regardless of strategy, a bull seal will be clapped out before turning twenty-five, whereas the cow can live for ten years longer.

As for the pups, after the yellow amniotic staining has washed away they're a beautiful creamy white. Couple that with the enormous dark eyes, and it's small wonder the RSPB features this unbirdlike creature on many of its marketing materials. Grey seal pups don't have waterproof fur, and it takes at least a fortnight for their lanugo (the first downy hair) to be replaced by a mature glossy coat. A harbour seal pup, on the other hand, is born with its dark adult coat already developed, which is why it can swim immediately; its first moult happens in the womb. A young grey seal pup is therefore unmistakable, and scientists speculate that their startling whiteness is a leftover from a time when the species bred on ice – as still happens at their Baltic outpost.

*

Watching a colony of seals is never dull: they honk, they bark, they yelp, they wail, and occasionally they grump and growl at each other as they negotiate their own private space or jostle for dominance. They like to lie on small rocks just above water level and, as the tide comes in and they can't bear to move, they bend their heads and tails upwards to perform the 'happy banana' shape. They're wary of humans on land, but in the sea they become much bolder. A few years ago, when our daughters were competing in the Devon Open Water championships at Oddicombe, a pitch invader in the form of a grey seal kept surfacing between swimmers. Open water swimming has little to recommend it as a spectator sport: parents peer through binoculars and argue over which identically coloured distant swimming cap belongs to their own darling offspring. So this improved the entertainment no end. One or two parents vaguely wondered whether their children were about to get eaten, but the swimmers themselves ploughed on regardless.

We've swum with seals off Lundy. You take a boat from Ilfracombe, drop anchor in a bay some distance from where seals are already hauled out on the rocks, wiggle into a wetsuit, jump overboard, and float around randomly for a bit. The rule is never to feed them and never to approach them; you mustn't do anything to make a seal move involuntarily. Let them come to you if they want, and they probably will sooner or later because they can resist everything except temptation. Seals love to play, and their favourite game is grandmother's flippersteps; they'll sneak up behind, wait until you're about to turn round, and then dive down underneath and come back up on the other side. Occasionally they can be a bit frisky, gently nibbling people's bottoms as they pass below. They're intrigued by the neoprene wetsuit and want to work out whether it's an integral part of the strange cumbersome creature that's thrashing around so gracelessly in the water. Fiona was on the receiving end of this treatment, which prompted our boat's skipper to tell us that they tend to nibble the women more than the men. We've no idea what to do with that particular piece of information.

At Lundy we saw about thirty grey seals, and we'd seen even more on the Eastern Isles of Scilly during our hedgehog hunt a month earlier. Those numbers still don't come close to the enormous colony sizes that you find along the east coast of England and Scotland. It's a long old poke from Devon, but finally we managed to clear our diaries for a weekend in the village of Horsey, on the coast about twenty miles north-east of Norwich. (The more frivolous of our readers may be amused by the name, but there's no equestrian of us foaling around with silly puns.) Horsey is home to thousands of grey seals during a breeding season that runs from November through to January or February. As one suitably impressed *Telegraph* article about Norfolk seals asked, 'Who needs the Serengeti?'

We both grew up in the west of England (albeit 300 miles apart), so for us Norfolk possesses an alien beauty. We love it, but we don't visit often enough to understand its landscapes. Frankly, we're tourists in our own country. Driving towards Horsey, we notice an occasional windmill interrupting the wide horizons, and we think inevitably of the Netherlands – to which a land bridge existed until about 9,000 years ago when the glaciers melted and the seas swallowed up what had been rich farmland. Since then, both the Dutch and the East Anglians have spent a lot of energy trying to protect themselves from further watery infractions, not always successfully. A storm surge in 1953 became the twentieth century's biggest natural disaster for both nations, killing over 300 people in Britain and nearly 2,000 Dutch, as well as many hundreds more on ships at sea.

The exposed eastern coastline can prove deadly for seals, too. Even in mild years many pups die, ripped off the beaches by clawing waves. If a serious storm coincides with high tides, the result is catastrophic. That point was brutally made just a week after our November visit, when Storm Arwen struck the east coast. The Scottish Borders bore the brunt; at one site near Berwick, well over 200 seal pups were killed. Horsey was far enough south to avoid the worst of the impact; nevertheless, by Christmas somewhere in the region of 100 pups had been lost. With no adult waterproof coat and insufficient reserves of fat, they don't last long in cold seas.

Horsey's biggest disaster of recent years came in December 2013, when the coastline was deluged by a tidal surge. At Hemsby, four miles south, seven clifftop homes and a lifeboat station tumbled into the sea. Horsey at that time had a much smaller breeding colony than today, but still more than 300 pups perished, their bodies washed up along the coast at Winterton. Survivors were raised by heroic volunteers, although the scale of that challenge can be measured by the respective weaning times: a grey seal fed on its mother's thick fatty milk more than trebles its birthweight in just three weeks; a rescued pup takes four to six months in a sanctuary, where it's brought up on a fish soup because all concoctions attempting to replicate seal milk induce diarrhoea.

Going on a seal expedition is crucially different from going in search of pine martens or wild boar. At the right time of year, you travel in certainty rather than hope. You *will* see seals, and the only question is how many. If you're visiting Horsey between late November and the end of January, the answer is hundreds, maybe thousands. There'll be male and female adults, and endless pups. Bizarrely, a colony of seals is called a rookery, and these rookeries are blobbed along the east coast with population sizes that make Horsey seem quite modest. An hour's drive north-west there's Blakeney Point, home to 4,000 pups in recent years; further north in Lincolnshire, near the southern mouth of the Humber estuary, 2,000 grey seal pups are born at Donna Nook, and another 2,000 are born in and around the Farne Islands. And so on, all the way up to Orkney and Shetland: more than eighty percent of our grey seals breed in Scotland. The pupping season, for reasons that aren't entirely clear, progresses in a clockwise direction around the British Isles. There are records of grey seal births in Britain for every month of the year, but the earliest births usually occur in the south-west, with Lundy's pupping in full swing during August and September. In northern Scotland it's September to November, and things get underway later on the south-east coast of England, where pups are still being born in February.

Done right, wildlife tourism should be not just sustainable but beneficial. If wildlife is more valuable to local communities alive than dead, it stands a far better chance of prospering. That's obviously true in Africa and South America, but even on holiday in Britain we always tell everyone why we're there. (Admittedly, we did that in the Forest of Dean when we were searching for wild boar, and got some very odd looks.) Some of the earliest protesters against seal culls in Britain were boatmen on the east coast who supplemented their regular income by taking visitors out to see the breeding colonies along inaccessible shorelines. The downside, of course, is that tourism risks destroying the very thing it wants to see. Close to 100,000 visitors come to Horsey for the seals each year and, judging by the busyness of the car park when we arrived, a significant proportion of them had chosen the same Saturday as us. Different sites have found different solutions to the problem of protecting the seals from disturbance. At Blakeney Point, you can't get near the seals by land because the beaches are closed during the breeding season, so the trick is to book a boat tour and approach from the sea instead.

It's ironic that the damage wrought by the storm in 2013 has allowed a closer engagement at Horsey. Dunes were swept away, leaving a flat and exposed beach. The remaining dunes are about 150 metres back from the shore – depending on the tide – and they've become the perfect viewing platform from which to gaze along the coast for a mile or so in either direction. The seals are scattered haphazardly on the sand below, unfussed by the human hordes. You couldn't design it better if you tried. Making sure that no one is tempted to go down onto the beach, a team of voluntary wardens from Friends of Horsey Seals patrols the paths; they keep an eye on the seals too, especially if there are concerns that a pup has been abandoned. Their Portakabin in the car park serves as an information point, and a nearby blackboard gives the latest census figures, compiled each week by volunteers wandering along the dunes with clickers. We'd been told to expect low numbers because it was still early in the season, and that year the seals were a fortnight behind in their breeding. So it was a happy surprise to see, written in large capitals, ADULTS 1663 PUPS 447. That short stretch of beach was currently home to a smidgen under 0.7% of the world's grey seals, and we were about to wander up and over the dunes to see them.

The same Portakabin prominently displays the true story of Mrs Frisbee, although our own detailed research (in the form of a five-minute Wiki Walk) reveals that she should have been called Mrs Aerobie. An aerobie is a flying ring: a frisbee with no middle. This particular seal had ended up with a ring wedged round her neck, and, as she grew, the plastic cut into her skin, gouging a wound that was only getting deeper. It seems particularly harsh to name her after the thing that was slowly trying to kill her: it'd be like calling Tim Mr Bowel Cancer. She suffered with that noose tightening for six months but, finally, close to death, she was captured and the ring was cut away. One long period of rehabilitation later, she was released back into the sea nearly three times heavier than when she'd first been rescued.

That tale is a warning about the risk to wildlife from discarded plastic, and it comes with a kicker: by 2050, the plastic in the oceans is predicted to outweigh the fish. Much of it is so-called 'ghost gear' – lost or deliberately discarded fishing nets. These make deadly traps for marine life, and grey seals are among the most common victims. The full extent of fatalities from ghost gear can't be known, but it's likely to be significant if evidence among living seals is any measure. Analysis of seals from a haul-out site in the south-west of England found that around four percent of animals every year had been (or still were) entangled in netting.[1] Of those, sixty-four percent had injuries that were deemed serious. This toll comes in addition to the estimated 488 seals that were killed by British boats as what is delicately called 'bycatch' in UK fisheries during 2019 alone; we have no statistics for other nationalities fishing those same waters, so the true total is likely to be at least twice as high. Although SMRU considers that these death rates are unlikely to damage national populations, they draw attention to the potential for an unsustainable decline in some colonies locally where gillnet fishing is prevalent (mainly in the south-west), with deaths so far being offset by immigration from other regions.[2] Many fishermen are at the forefront of innovations designed to avoid or reduce fatalities; others continue to dump worn-out gillnets in the seas because they can't be bothered to dispose of them responsibly. Some trawlers carry up to seventy miles of nets, so just one incident can bring about massive destruction to marine life. Governments need to start rewarding the blameless crews and cracking down on the cowboys.

Just before we peek over the dunes and gasp at the seals below, let's credit some of the other reasons for wanting to visit the coast at Horsey. The dunes from Horsey down to Winterton are designated as a Special Area of Conservation, comprising, we're told, 'the only significant area of dune heath on the east coast of England'. This makes them home to a variety of rare and not-so-rare grasses. They're anchored by marram grass, which won't get a botanist especially excited. Nor will the Yorkshire fog, but the grey hair-grass may. We'd better confess that, philistines as we are, we struggle to get worked up about grasses, whether they're rare or not. The wildlife that they attract to the dunes and heaths is another matter. This is an important site for butterflies like the grayling and the dark green fritillary, not that you'd know it in November. The birds are there in the winter; without having to look especially hard, we saw stonechats, sanderlings, and all sorts of gulls. Kestrels working the dunes were a sure sign that the area was rich with small mammal activity. We have a soft spot for kestrels, having twice rescued and rehabilitated juveniles who'd been knocked out of the sky by disapproving corvids. We're not going to stand by and do nothing while a grounded kestrel has its eyes pecked out by a murder of crows.

Best of all – because we rarely see them back in Devon – were the pink-footed geese that had congregated in large numbers about half a mile away. Like the grey seal, this species relies heavily on Britain for its home: more than eighty percent of the world's population overwinters here. A pink-footed goose that ends up in Norfolk will probably have started its journey in Iceland or Greenland, heading south as the weather turns chilly. It might stop off in Lancashire, timing its arrival with the carrot and potato harvest so that it can feast on all the leftovers.

Arrived at its Norfolk destination, it carries out a similar role, this time tucking into sugar beet tops. The farmers don't mind at all; while it's padding around in the fields eating remnants, it's also fertilising the land. Partly because of these agricultural cycles, the pink-footed goose's numbers in Britain have soared more than tenfold, from 40,000 in 1960 to 490,000 today. We reckon that at least 500 of them were at Horsey, and something – maybe a fox – must have spooked them because with perfect timing they took to the skies and performed a lengthy and spectacular goosuration for our benefit. (That's our younger daughter's word; don't write in!)

A seal-crowded beach in late November has more drama than an *EastEnders* Christmas Special. Bulls have already begun fighting for territories, because cows come into oestrus only a fortnight after giving birth. Beforehand, the females aren't keen to let the males anywhere near them or their pups; they show a fine set of teeth to any male who ventures too close. It's fair enough because they have other priorities: they're undergoing a cheap and effective liposuction courtesy of their offspring, losing up to ten pounds in weight every day while lactating. Meanwhile, the pup is vulnerable to the clumsy 600-pound sumos thrashing around them. We're grateful that we've never seen it happen, but occasionally a pup does get crushed in the uproar. For one reason or another, between a fifth and a quarter of grey seal pups die in the first three weeks. At Horsey there were several dead pups on the beach, although whether they'd been squashed, stillborn, or swept into the sea and then washed up wasn't obvious without a closer inspection. The opportunistic gulls didn't agonise over the niceties as they made short work of their seasonal manna.

Gulls are more than bit-part actors in any seal-watching. The mother seal tends to give birth at night on the beach, but, having pushed a thirty-five-pound pup out through her birth canal, she's in no hurry to move for a few days. So she lies there, next to a pup stained yellow from the amniotic fluid and with its umbilicus still attached. The afterbirth is close by – tantalisingly close – and gulls consider seal afterbirth a beak-smacking delicacy. The mother seal will tolerate the gulls' presence only up to a point, so what follows is a stand-off as they wait impatiently to bury their beaks deep into that lovely red goo. When the mother finally shuffles far enough away, the beach comes alive with noisy and excited birds. We watched a single great black-backed gull standing on a placenta and holding out against a circle of herring gulls. Every time our solitary hero took a peck at its prize, the herring gulls jumped in to rip at the edges. Here was the dilemma: when it ate, it was robbed; when it lifted its head and gave them a dirty look, the herring gulls retreated. That went on for an hour back and forth, before turning into a free-for-all amidst a blizzard of agitated wings.

One female seal was searching desperately for her baby. She tried to inspect every pup she spotted, often getting angrily driven away by the pup's actual mother. Still she kept going, calling all the time in her panic. We wanted the happy ending, but after a while it still hadn't arrived, and there was only a certain amount more that we could comfortably watch. The other side of the problem is just as common: pups can't find their mothers. At Horsey the wardens have a simple technique for monitoring them: any pup that they're concerned about is sprayed with blue paint so that they can keep an eye on it. They don't like going onto the beach because it worries the adults and flushes many of them into the sea, but they'll intervene if they absolutely must. Fostering of pups among grey seal populations isn't unheard of, but those pups don't seem to grow at a normal rate and most will probably die before adulthood.

Like it or not, all grey seal pups are abruptly abandoned when they're about eighteen days old. Having tripled the weight of her pup to around forty-five kilograms, and lost forty percent of her own body weight in the process, the mother seal decides that enough is enough and returns to the sea. It must be some kind of cruel evolutionary joke that females come into oestrus again at around the time that they stop feeding their pups; they don't get a moment's peace, with males lurking for the opportunity to mate either in the shallows or on the beach. This may all seem like indecent haste, but at least it makes the seals' love lives much easier to manage than if they had to find each other at sea after dispersing: individuals can travel hundreds of miles from maternity colonies within a fortnight.[3] Left high and dry by these parental frolickings, the pup moults and then eats nothing for the next three weeks. Perhaps hunger is a necessary stimulus to persuade it finally to take the plunge and start hunting its food.

Foraging underwater to catch that food is a murky business; even when seas are relatively clear, the light doesn't penetrate far. At ten metres below the surface, ninety percent of light has been absorbed or reflected, and only one percent reaches forty metres down. Seals need extremely large eyes for their body size. They're proud owners of a tapetum – a layer of reflective tissue that passes light back through the sensory cells a second time. (The eyeshine of an animal that you spot in your headlights as you're driving at night is a reflection from the tapetum.) Seals are also much better than us at hearing underwater, thanks to hugely enlarged ossicles – the bones of the inner ear. We think that they use smell to track chemicals such as dimethyl sulphide, which is released when zooplankton feed on phytoplankton; those are the areas likely to be richest in fish. Then there are the most astonishing sensory organs of all: whiskers. With sensitivity far greater than the human fingertip and crammed with ten times more nerves than the whiskers of other mammals, these can detect the hydrodynamic trails left by fish swimming more than 180 metres away.

Fiona is one of those crazy all-year-round no-wetsuit sea swimmers whom people point at from the safety of the shore, so she can testify that her fingers go numb within seconds of being submerged in cold seawater. Her blood is drawn away from her body's surface towards the core, which causes muscles to tighten and nerves at the extremities to shut down. The seal suffers none of this. It's generally adapted to immersion in a cold liquid, with a thick layer of blubber and a rounded shape that gives a better surface area to volume ratio for minimising heat loss. The seal operates a counter-current system that keeps the base of the whiskers toasty, the nerves fully functioning and the muscles mobile. So effective is this internal heating that individual follicles show up clearly on photographs taken with a thermal imaging camera.

Seals can detect objects just a millimetre or so in diameter using their whiskers. Even more impressively, they don't have to touch them at all. Because water isn't compressible, the vibrations caused by the tiny whirlpools generated in the wake of swimming animals are transmitted directly to the whiskers. When hunting, seals stick their whiskers out in front as a kind of antenna. Their ability to sense creatures moving 2,000 times more slowly, amidst a general whooshing of water all over their bodies, intrigues fluid dynamicists and evolutionary biologists alike. The whiskers have a unique three-dimensional wavy structure that cancels out the vibrations created by the seal's own movement. Equipped with these high-performance detectors, seals can work out the speed, direction and size of any fish that was passing twenty seconds earlier.[4] Sight isn't strictly necessary. There are reliable reports of a grey seal in the Farne Islands successfully raising pups three years running, despite being completely blind.

The grey seal was the first British mammal to be protected by an Act of Parliament – a dubious honour prompted by its having been persecuted almost to extinction. Because it spends so much time out on land during the breeding season, and pups left behind by their mothers are defenceless, the species is vulnerable to human hunters. The very few places where it thrived in the nineteenth century tended to be remote and uninhabited Scottish islands. By 1914, some estimates put the numbers of grey seals in British waters as low as 500 individuals, so the Grey Seals (Protection) Act of that year banned the killing of seals during a breeding season that ran from 1 October to 15 December. In 1932, an amendment stretched the period in both directions, from the start of September to the end of December. That didn't stop almost every seal pup born in Shetland from continuing to be hunted and killed, but, despite sounding rather unambitious in its scope, the law seems to have rescued the species: a survey in 1966 estimated a total of 35,000 grey seals, and that number had almost trebled by the late 1980s. The increase came thanks to further help in 1970 from the Conservation of Seals Act, which extended protection to include the harbour seal. Even so, the legislation contained as many holes as a string vest. You weren't allowed to kill a seal deliberately – unless, that is, you believed it was going to 'damage fisheries', or you wanted to bring about 'the reduction of a population surplus of seals for management purposes', or you were keen to 'use a population surplus of seals as a resource'. Rarely have euphemisms sounded so sinister. The law was mealy-mouthed enough to allow the National Trust to continue culling well over a thousand seals – males, females and pups – on the Farne Islands throughout the early 1970s, and there were significant culls in Scotland later in the decade before public opinion brought a halt to the massacres. Scotland now has its own Marine (Scotland) Act 2010 that supersedes the British legislation but duplicates many of its weaknesses.

The killings didn't stop in the 1970s; they became clandestine. A new licensing system was introduced that may have been intended to protect seals, but it ended up giving persecution a veneer of legality and respectability. In Scotland under its new Act, licences were granted for the shooting of 1,025 grey seals and 314 harbour seals in 2011 alone, the major culprits being industrial salmon farms. Official figures show that there were fewer seals shot than licences granted, but what sounds like good news wasn't anything of the kind. Not everyone will have waited for the licence to arrive, or even bothered applying for it in the first place. It remained perfectly legal to shoot seals via the fisheries defence clause, and a single licence could grant permission to kill multiple animals.

The Scottish government's loyalties have been brazenly clear throughout. We normally think of politicians as our lizard overlords, but Nicola Sturgeon is named after a fish, and Alec Salmond was a near miss. As governments dictate the policies of the Statutory Nature Conservation Organisations, salvation needed to come from an external source: in this case, surprisingly, the Americans. In 2016, the US government passed a new animal welfare law (the catchily named US Marine Mammal Protection Act Import Provision Rule) banning the import of fish from countries that permit the culling of seals to protect fisheries. During the five-year delay before it was implemented, the Scottish government lobbied Washington to be granted an exemption. When that failed, they reluctantly changed the law to comply with US legislation and avoid jeopardising the £180 million annual trade to the USA in farmed salmon. T.S. Eliot once wrote about 'the greatest treason: / To do the right deed for the wrong reason.' A profit-hungry government forced against its will to make an ethical decision may have been what he had in mind.

The Scottish government aims to double the size of the salmon farming industry by 2030, and until those meddling Yanks came along it saw no reason why seals should be allowed to inconvenience its plans. The Scottish National Party's Fisheries spokesman had called for a renewed seal cull as long ago as 1999, arguing that jobs should not be sacrificed on what he unimaginatively termed the 'altar of political correctness'. (The spokesman in question was Fergus Ewing, from whom we'll be hearing more in due course.) The growth of salmon farming since then has only exacerbated the tensions. Politicians are now heavily invested in an unsustainable industry.

Back in the early 1970s, someone in Norway came up with the idea of producing Atlantic salmon in huge floating cages. Delivering a gruesomely high yield of eight kilograms of salmon per cubic metre – which really tells you everything you need to know – the farms ensured that salmon went from being a rare treat for high days and holidays to something readily available at the corner store. More than 200,000 tonnes of the stuff is now produced in Scotland every year, making farmed salmon the UK's largest food export product by value. This sounds like a good news story, and squinting at it from a particular angle you might persuade yourself that salmon farming is the answer to concerns about food security and overfishing. Yet the welfare issues are so intractable, and the environmental damage so extensive, that it's hard to see how any benefits could justify the costs.

Salmon are migratory and carnivorous. Farmed specimens are far less energy-efficient to produce than any herbivorous species; they provide food with lower nutrient densities than the fish that are caught to feed them. One study showed that in 2014 alone, 460,000 tonnes of wild-caught fish was used to yield 179,000 tonnes of Scottish farmed salmon.[5] Over three-quarters of that catch comprised fish species commonly eaten by people – so much for addressing food security – and the remaining species, such as sand eels, are vital to marine ecosystems. Some of these issues have been partially addressed: the industry is trying to turn its salmon vegetarian, and only about one-fifth of the feed is now derived from fish. But salmon need protein, and much of the deficit is provided by soy – a crop that's heavily implicated in deforestation and soil erosion. About ninety percent of all the soy harvested in the world is fed to livestock, including salmon.

Living in such tight proximity, farmed salmon have been described as a 'Petri dish' for all sorts of parasites and pathogens that can destroy wild populations. Sea lice, for example, skin the flesh from the salmon's head, and are quite capable of devastating a pen of 200,000 fish. Just as the sea lice escape into the wild and cause serious damage, so do the salmon themselves. Wild salmon around fish farms now carry the genetic signature of domestic stock, the result of interbreeding between female escapees and native males. These fish are simply not fitted to their environment, and they die off in vast numbers;[6] it's yet another reason why wild salmon populations have halved since salmon farming began. According to official Scottish government figures, more than 200,000 farmed salmon escaped in 2020, and more than 3.5 million have escaped since 1998. The RSPCA nevertheless gives its blessing to the industry, stamping its 'Assured' label on farms meeting certain fairly modest welfare criteria; more than sixty percent of Scottish salmon farms are now signed up to the scheme, which, until the Americans stepped in, allowed marksmen to shoot seals as a 'last resort'. If the RSPCA ever doubted its own integrity in this matter, the more than half a million pounds that it receives from the industry every year must have soothed its troubled brow.

The most conscientious salmon farms were turning to non-lethal methods of dealing with seals even before that American intervention. Over eighty percent of fish farms in the Western Isles, Argyll and Bute and the Highlands have installed acoustic deterrent devices (ADDs) to try to scare seals off, but some seals have learned to recognise the sound as a dinner bell,[7] and others become habituated and stop taking any notice. ADDs have significant downsides: they disturb cetaceans and have even induced hearing loss in harbour porpoises. So producers have started investigating better solutions, such as rigid reinforced mesh for cages. This has the advantage of being difficult for seals to chew through; holes made by seals are a major cause of farmed salmon escapes. It also prevents the seals from adopting the clever strategy of catching a fish in a pocket created by a deformed piece of netting, biting it in half, and sucking all those delicious guts out through the net. Tensioned anti-predator nets are being deployed as an extra layer of defence around the cages to stop the seals getting anywhere near. These avoid the risk of fish being spooked by circling predators, which is bad for fish welfare and stunts their growth rates. A few farms have added top nets to prevent the most intrepid seals from climbing into pens and swimming with their supper. All of these are vast improvements on the ruthless methods employed previously, but anti-predator nets have to be installed with care because otherwise they entangle and drown seals and other marine animals.

If you bake a cake and invite us round to eat a slice, you end up with less cake. On the same principle, if a seal eats a fish there's one less fish for the fisherman to catch. Culls of seals in modern times have been motivated by this cake theory of resources. In 1977, an official cull got underway after the Department of Agriculture and Fisheries for Scotland picked some numbers out of the air to claim that annual fish consumption by seals in Scottish waters represented a loss to the fishing industry of 65,000 tonnes, with a market value of between £15 million and £20 million. Several years later and outfaced by vociferous public disapproval, the department commissioned a major study from SMRU to examine the relationship between seal numbers and fish stocks. The research concluded that a general reduction in seals would probably not affect commercial fish catches. That was forty years ago, and in all its subsequent work SMRU hasn't felt the need to waver. In 2010, an official European Parliament study of seals and Scottish fish found that 'the effects of predation by both species of seals on overall stock abundance of most fish species is likely to be insignificant', and that 'even a large reduction in the number of seals in Scottish waters would be unlikely to make any noticeable impact to the success of demersal or pelagic fisheries'. That's because ecosystems are more dynamic than cakes.

Seals take less than one percent of the North Sea's fish biomass but, even if it were much more, the differences to fisheries would be negligible. Imagine that seals and humans agree from now on to stop catching fish altogether; the fish that would have been eaten are able to breed, and their offspring breed, and their offspring's offspring breed, and before very many generations there isn't enough sea to fit all the fish in. (Media scare stories about rats provide a familiar variant: if not urgently controlled, they'll swarm in plague proportions and eat out our eyes while we're sleeping.) Newsflash: that doesn't happen. Animal populations have a natural carrying capacity – a maximum level that the environment can sustain. Go beyond that point and populations will naturally fall because of depressed breeding and survival. For example, the biggest predator of fish is other fish, and those predatorial species will thrive until the glut disappears. At the opposite end, populations below their carrying capacity will breed and survive better because there's a surfeit of resources. This pattern has a happy consequence for anyone wanting to harvest fish, whether seal or fisherman. Somewhere below the carrying capacity is a point – known as the maximum sustainable yield – which is the largest amount of fish that can be removed without negatively affecting the population. At the risk of mixing our food metaphors, we can think of this as the magic porridge pot scenario, where we harvest but the goods keep on coming. As long as we don't get greedy, compensatory reproduction and survival make up for the loss. This is the whole principle behind fishing quotas, and it's why neither the various mathematical models nor events in the real world have ever been able to demonstrate that seals adversely affect fishermen's livelihoods.

Despite those occasional unscientific culls, grey seal numbers have been increasing steadily around Britain since detailed surveys began in the late 1960s, and probably since 1914. While the populations in Orkney and Shetland are now levelling off, those in the North Sea and particularly in southern regions such as Horsey continue to grow rapidly. The changes are largely the result of increased pup production, but there's also some immigration; having travelled to the foraging grounds of the southern North Sea, some mother seals decide to stay there rather than return to their birthplace.[8] The same regional shift seems to hold for other marine species such as harbour porpoises, now at their highest densities much further south than their 1990s strongholds in the northern North Sea and Orkney.

Harbour seals are also doing best in southern regions, but they were badly hit there by the phocine distemper outbreaks of 1998 and 2002. Although the cause was never adequately explained, various theories were posited involving algal blooms and food depletion. The impacts on population sizes were also puzzling. In the south-east, the two outbreaks were apocalyptic, wiping out fifty percent and thirty percent respectively of the harbour seal population, but elsewhere their impact was much lower. Even aside from those epidemics, there continues to be a nagging worry that harbour seals may be in long-term decline. Recent analyses show that their numbers are similar to those in the 1990s, but if that makes us complacent we're probably victims of our old friend Shifting Baseline Syndrome: populations at that time were artificially low, having not yet fully recovered from the previous decades of persecution. Numbers of harbour seals aren't uniformly stable; they've declined substantially in some regions and increased in others. Populations in the west of Scotland and the Western Isles have remained constant, whereas those in the east (including Orkney and Shetland) have collapsed. Further down into the north-east of England harbour seals are doing well, as they were in the south-east until 2019 when there was a sudden and unexpected fall of twenty-seven percent, with subsequent local surveys confirming that the decline continues.

These patterns are probably influenced by a range of factors, but harbour seal numbers have fallen furthest and fastest where sand eels have been scooped out of the North Sea in their billions. If you're not sure what sand eels are, think of the photos of puffins with beaks stuffed with a row of tiny silver fish. A sand eel isn't an eel at all, but a generic term for a number of small fish species that help to make marine ecosystems function. Sand eels are harvested on an unimaginable scale by industrial trawlers, and, when they're not being ground into pellets for farmed salmon, they can be used for fertiliser, burned to generate electricity, or added to cattle feed – which sums up quite neatly everything that's wrong with the world. The next time your dog on its countryside walk decides to sneak off and start tucking into a cowpat, the chances are that it's attracted by the stink of fish.

*

Could the increase in grey seal numbers be having a negative effect on harbour seals? Grey seals pack away between four and seven kilograms of cod, haddock, ling, whiting, sand eel and flatfish every day, with a few octopus and squid thrown in for dessert. By virtue of their size, they're undoubtedly dominant over their smaller cousins and likely to outcompete them when resources are sparse, but the two can be seen living side by side in many parts of our coastline. Hidden among those 2,000 grey seals when we visited Horsey was a single harbour seal, minding its own business and remarkably relaxed about being the odd one out.

In case you hadn't noticed, we're fans of grey seals. They're inquisitive, charismatic, characterful and (although not exactly pretty) very handsome. But we grudgingly admit that a small number seem to have acquired less-than-endearing habits, such as snacking on harbour seals and harbour porpoises, and cannibalising pups of their own species. This probably started innocuously, with hungry seals scavenging the carcasses of dead pups and other marine animals drowned by fishing gear. Then they developed a taste for it. Reports of attacks on marine mammals by adult male grey seals have been increasing for about a decade; we don't know whether this Lecter-like behaviour is spreading or whether we've become more attuned to noticing it. Either way, the total amount of predation hasn't been quantified, and SMRU is currently asking the public to submit photos of grey seals that they've seen attacking other marine mammals. The plan is to create a rogues' gallery and work out how many animals are guilty, identifying the wrong'uns on the basis of unique markings.

Effects on population sizes of harbour seals at national or even regional levels are expected to be minimal. Nevertheless, greys have been killing notable numbers of harbour seals locally at Blakeney and in Heligoland off the coast of Germany. In Orkney, where grey seal populations remain stable, harbour seal numbers have fallen by eighty-five percent since the 1990s. If just one percent of the adult male grey seals in the area killed six harbour seals per year, that would fully account for the decline.[9] Under interrogation, grey seals would doubtless implicate the orcas that are well-known predators of harbour seals. Again, a small number of orcas eating a lot of seals would be enough to explain the falling numbers,[10] but little is currently known about predation rates, so at the moment we can only guess at the causes of the harbour seal's patchy declines.

We hope that the following flippering fantastic facts will seal your enthusiasm for the species:

There's some disagreement over whether the ancestors of seals are best described as otter-like or bear-like. Genetic research tells us that seals are related to both. Pinnipeds didn't go straight from land to sea; there was an interim stage during which they hunted in fresh water. A fossil specimen discovered in 2007, *Puijila darwini*, has been described as a walking seal, with a seal's skull and teeth but no flippers.

There are three genetically distinct populations of grey seals that diverged around a million years ago:

the western Atlantic seals on the coast of Canada and down to New England; the Eastern Atlantic population on the shores of Britain and Ireland, with a few more scattered across Iceland, the Scandinavian coast, and along the North Sea's continental shores; and a much smaller population in the Baltic Sea. Estimates are wildly divergent, but the IUCN in 2016 gave a grand total of 316,000 mature grey seals globally.

Strictly speaking, earless seals aren't earless but ear-flapless. Although only eighteen out of the thirty-four pinniped species are phocids, they make up roughly ninety percent of all the pinnipeds in the world, and can be found in both the Arctic and Antarctic regions and in many seas between. When Aristotle wasn't insulting the seal's gait, he took exception to its earlessness: 'One viviparous animal, the seal, has no ears but only auditory passages; but this is because, though a quadruped, it is deformed.' Seals prefer Plato.

Males become sexually mature at six to ten years, whereas female grey seals can start to breed at three to five years old. Gestation lasts about eight months. To make sure that the breeding cycle takes a year, implantation of the fertilised egg is delayed by four months after mating.

Aside from their extensive repertoire of grunts and growls and barks, grey seals have a few non-vocal techniques for communicating. They create underwater claps by knocking their flippers together, and individuals at Donna Nook have taken to performing body slams worthy of Big Daddy: large males impress rivals by lifting their torso up with their front flippers and letting it slap noisily to the ground. The resulting vibrations give potential opponents a good indication of their body size, and probably help avoid unnecessary fights. First reported in 2014, this behaviour seems to be confined to the east coast.

Grey seals spend about eighty percent of their lives at sea, and most of that (around ninety percent) is underwater. They prefer shallows and make short dives of between three and ten minutes to forage, but they're capable of diving for half an hour without needing to surface for more oxygen. They can even take naps on the seabed. All of this is owing to an extraordinary physiological trick: on a long dive, they reduce their heart rate to just four or five beats per minute; when they're above water it's as much as thirty times higher, at about 120 beats per minute.

The Montgaudier baton is one of the most extraordinary pieces of prehistoric art ever discovered. It's a thirty-centimetre reindeer antler on which has been engraved, in impressive detail, two grey seals chasing a fish. The baton was found in a Dordogne cave well over 100 miles from the sea. Seal bones have also been uncovered in the vicinity, prompting the hypothesis that seals were routinely following salmon or trout upriver for vast distances. Google an image of the Montgaudier baton and you'll notice that the fish seems to be swimming upside down. Some seal experts have argued that, in fact, the baton is being displayed the wrong way round: it's not the fish but the seals that should be upside down. In a river, there'll be no prey or predators above them; because of the position of their eyes, they swim on their backs to get a clear view of the bed. One estimate, based on the behaviour of seals at the New England Aquarium in the USA, is that they swim upside down about seventy percent of the time. This behaviour is also common in the wild.

Grey seals regularly demonstrate a propensity to show up in unlikely places. Fiona has seen one in the Thames at Millbank, and they're spotted in the Tyne at Newcastle. In 2016, a grey seal caused a local sensation when it somehow reached the River Tone near Taunton, about seventeen miles from the sea. That record was beaten in 2018 by a harbour seal on the River Swale in Thirsk and a grey seal on the River Wharfe near Tadcaster – both more than fifty miles from the coast.

In 1999, Derek Yalden worked out that among native or naturalised British mammal species, the grey seal came third in total biomass behind rabbits and red deer. Since then, rabbit and red deer numbers have fallen. Thanks to a fifty percent increase in population size over

> the past two decades, the grey seal has overtaken them both and is now top of the biomass league among wild British mammals. It still doesn't come anywhere near the numbers for horses, pigs, cattle or sheep – or, of course, humans.

The long winter nights in Orkney and Shetland can do funny things to a man, especially if he's a lonely bachelor. How else to explain the legend of the selkie? The Scots word for seal is 'sealgh' or 'selch', and 'selkie' is the diminutive; it describes a creature with the appearance of a seal and the ability to change to human form and back again. In the story's most common version, this transformation happens when the seal sheds her skin – or, more usually, has it stolen by a man. Condemned to a life on land, the selkie then marries the thief and they start a family; all the while she pines for her previous existence. One day she finds her skin, and the lure of the sea proves greater than her love for her family. She slips into the water beneath the waves, never to be seen again. The legend encapsulates the alluringly human nature of the seal. Its large expressive eyes hint at depths of soul more profound than any other non-primate's. Dolphins and whales have left us behind, but seals are go-betweens exploring the full extent of our terraqueous globe, and on that thin shoreline interface they seem to possess secrets that we can only glimpse.

It may be its vast vocal range that confirms the seal's special status. Mother seals can recognise the call of their offspring amidst the hubbub of a rookery heaving with pups. The eerie howl coming across the waters from colonies of grey seals has frightened many a sailor half to death, and helped to popularise the belief that seals are souls of the drowned. In captivity, grey seals have been trained to reproduce parts of 'Twinkle, Twinkle, Little Star' and the *Star Wars* theme tune. They're not bad: you wouldn't want them in your boy band, but it's obvious what they're performing. Some enterprising scientists in search of headlines have claimed that the sirens of Homer's *Odyssey*, whose beautiful singing lured sailors to their deaths, were probably Mediterranean monk seals. Monk seals were once common but are now listed among the dozen most endangered mammal species globally; it turns out that these particular sirens were nothing like as deadly as the sailors they were enticing ashore.

Female grey seals are often faithful to their traditional pupping sites, returning to within metres of where they gave birth the previous season. Many pupping sites across Europe are designated as Special Areas of Conservation (SACs) under the EU's Habitats Directive, which is still in force across Britain even after Brexit. SACs provide legislative protection against harassment of seals, and there's a requirement that measures will be put in place to ensure 'the integrity of the SAC', meaning that animals should be able to migrate to and fro without hindrance.

Grey seals also come ashore to moult. Females shed their two layers of dense waterproof fur between January and March; the males wait until later in the spring. They don't eat during the six-week process of growing their new coat, because fur production requires increased blood supply to the skin and going in the sea could therefore lead to catastrophic heat loss. Outside the moulting and breeding season, seals also need haul-out sites to rest and digest food. The physiological changes required for deep dives are incompatible with digestion, so grey seals delay the processing of food until they're resting.[11] Usually this takes place on land, but, given that foraging trips can cover seventy-five to a hundred kilometres per day, around ten percent of their time is spent resting offshore, saving energy on travelling back and forth.[12]

To protect a rare plant, lichen or butterfly, you put a fence round the place where it's found and declare it a nature reserve. A seal, on the other hand, is highly mobile and reliant on multiple sites for different activities. Researchers in Cornwall have worked out that some of their seals use over fifty locations, from the Isles of Scilly to Pembrokeshire and northern France. Of these, only a tiny percentage are protected by SAC designation. The lack of official status means that, as the Cornish monitoring shows, those sites are vulnerable to disturbance all year round. Walkers, RIB riders, kayakers, paddleboarders, swimmers, drone-flyers, people taking selfies – all have caused disruption to seals that should have been left peacefully digesting fish and dreaming of their childhood days. There's also been a significant increase in disturbance over time, probably boosted by the upturn in outdoor sports and pandemic staycations. Lately, the stories have been coming in thick and fast from across the country: the fatal dog attack on Freddie the harbour seal near Hammersmith Bridge; the stampede of 100 grey seals and their nursing pups at Robin Hood's Bay in North Yorkshire as an idiot runner deliberately drove them into the sea; the capture and relocation from Plymouth to Scotland of Spearmint the seal after she'd become habituated to the locals who regularly fed her. The excellent 'Give Seals Space' campaign is now raising public awareness, and 'Watching Seals Well' signs are on display around our coastlines with advice on how to behave responsibly.

Legislative changes are also on the way. While the Wildlife and Countryside Act prevents deliberate killing of seals, it doesn't preclude disturbance or harassment. That anomaly is odd because the protection already applies to other species such as whales and dolphins. One of the changes proposed by the Mammal Society – and now accepted by the Joint Nature Conservation Committee in its advice to ministers – is that the government should award seals full protection. The MP Tracey Crouch is proposing the same amendment via a Private Members' Bill, so, with any luck, the law will soon be tightened up.

What about when seals aren't on land? The world's seas are becoming increasingly dangerous, and the anthropogenic racket from shipping, wind turbines and piledriving is getting louder all the time. Piledriving for wind farms and gas rigs produces short bursts of extremely loud underwater noise that seals, funnily enough, don't appear to appreciate. Better news is that the noise from operational turbines isn't a problem for them. If anything, wind farms offer good feeding opportunities: trawlers are excluded and the basal structures act as artificial reefs where fish can spawn and grow in peace. It may be that the sudden spike in grey seal numbers at Horsey was prompted by the building of an offshore wind farm in the North Sea. We can't sound so positive about the construction of tidal-powered generators. Their underwater turbines pulp fish, and recent research shows that seals as far as two kilometres away are adversely affected by their noise.[13] Many developments have been proposed for locations where harbour seals are already in decline, and not much is being done to mitigate the potential damage to mammals at sea.

The UK government proudly declares that almost a third of our territorial waters are designated as Marine Protected Areas (MPAs). Confusingly, MPA is an umbrella term applied to different kinds of site: marine nature reserves like the very first one at Lundy; sites designated under European legislation as Special Areas of Conservation or Special Protection Areas; and, newest of all, Marine Conservation Zones (MCZs, known in Scotland as Nature Conservation Marine Protected Areas). All of this is a mouthful because the policy is a mess. The first MCZs – sorry about the clutter of acronyms – were introduced in 2013, and there are now more than 200 of them, with different criteria and management priorities in each of the devolved nations. Unlike most other designations, MCZs take into account social and economic considerations, as well as (and sometimes instead of) conservation objectives. They've been blasted by experts and marine conservation organisations as 'paper parks' for appearing to provide protection but delivering very little in practice. In 2021, sixty-two of the sixty-four MPAs that were meant to protect life on the seabed were still being spoiled by dredging and bottom-trawling. Frustrated by this lack of progress, Greenpeace has taken matters into its own hands and started dropping concrete blocks into the sea.

We're fortunate to have one of the more effective MPAs almost on our doorstep. The delicate reef system of Lyme Bay had been damaged by years of dredging for scallops. A partnership project with local fisheries and other interests has almost completely eliminated bottom-towed gear, and research by the University of Plymouth reveals that the area now boasts nearly four times as many fish – and more than four times as many *species* of fish – as adjacent areas where bottom-towed fishing is still allowed.[14] One result is the recent opening on our own street of an outlet selling fresh fish that has been caught locally by sustainable methods. The visible success of the Lyme Bay MPA highlights the lack of progress elsewhere. If these designations are to prove more than a greenwashing exercise, the government needs to enforce the protection of our marine life from supertrawlers, bottom dredging, extraction and discarded fishing gear.

Our opening chapter told the most exciting mammal conservation story of recent decades: the return of the beaver to British waterways. This final chapter has gone further back in time to the biggest success of the last century, when populations of grey seals along our coastlines grew from 500 to well over 100,000 individuals. In both cases – beaver and seal – the politicians said no and the public overruled them. The grey seals at Horsey and all over the British Isles are a glorious legacy, showing what can be achieved when we hold administrations to account. The seals couldn't even rely on bodies like the National Trust or Nature Conservancy – whose first director-general dismissed them as 'tiresome organisms'. What allowed them to flourish was vocal public support. The environmental historian Rob Lambert has described how campaigns to save the seals mobilised three powerful groups: wildlife enthusiasts, animal welfare supporters and – critically – the people who 'on a weekend walk at the beach hope to see a seal, or just want to know that seals are out there doing well'.[15] Together, they form a constituency that politicians daren't ignore, whatever the lobbyists may be whispering in their ear.

Of course, it's easy to galvanise public opinion when a big-eyed seal pup is being shot or clubbed to death. Many of our generation will remember watching the reports on *John Craven's Newsround* and feeling utterly traumatised. When the British government finally caved in to pressure and called off a seal cull in October 1978, every major national newspaper led with the story. That success inspired later struggles – to ban fox hunting, to halt stag hunting on National Trust land, or to cancel the government's senseless cull of badgers. The task now is to channel public passion into an effective protest against the indirect ways in which we destroy animals' lives: the ripping out of hedges, the dousing of fields with chemicals, the destruction of seabeds, the overgrazing of national parks, and so on. These things happen on a daily basis without inspiring the same crystallised outrage that a seal cull instantly provokes.

You won't regret making the trip to Horsey or any of the other rookeries along our British coastlines. They provide the most extraordinary natural spectacle that our country can offer. As you gaze in wonder at the seals, remember to give thanks to all the protesters who fought to make it possible. There are many battles ahead, and the Horsey seals are proof that we can win.

AFTERWORD

Reintroducing Cats and Dogs

At an early stage, we'd intended to include a chapter on wildcats. They certainly qualified as endangered – in fact, Critically Endangered – because their estimated numbers had fallen from 3,500 in the 1995 review to just 200 individuals by 2018. Having become extinct in England and Wales in the nineteenth century, the species is sometimes known by default as the 'Scottish wildcat'; it survived patchily in the north-east of the country, usually around broadleaved woodland, where it hunted small mammals and birds. Our plan was to journey to its heartlands in Ross and Cromarty and see what we'd find. In the end, we decided not to bother. We knew we'd be wasting our time.

Conservationists declare a species 'functionally extinct' when its populations are so small that they're no longer viable. Nobody can say definitively that there isn't a single wildcat left; the last one may be dead already, or it may linger for a few years, but that makes no difference on a species level. Wildcats are now, to all intents and purposes, extinct in Britain. There was a last-ditch effort to rescue them by Scottish Wildcat Action, a conglomerate of conservation organisations working with NatureScot and funded with lottery money. They carried out monitoring with trail cameras to find the remaining animals, and introduced a scheme for neutering feral domestic cats. It was too little, too late. In 2020 they admitted defeat and switched their focus to developing a captive breeding programme, tacitly acknowledging that there were no wildcats left for them to save.

The wildcat's fate is a devastating reminder that all those warnings about imminent extinctions aren't crying wolf. Two native British mammals have effectively gone extinct in our lifetimes (the greater mouse-eared bat is the other), and we as a nation ought to feel ashamed. The wildcat fell victim to the usual pincer movement of habitat loss and direct persecution by humans, with an extra problem thrown in for good measure: at low densities, it breeds with feral cats and therefore hybridises itself out of existence. When a wild boar population is tainted with domestic pig genes, those genes quickly disappear under selection pressures if they're ill-adapted to the environment. The same doesn't happen with wildcats, because the feral cat population benefits from human support and is constantly getting replenished from the ever-expanding pool of domestic cats. Hybridisation is therefore dilution and deterioration – away from pure wildcat and towards a semi-feral moggy. As if that weren't bad enough, wildcats have been trapped and shot by gamekeepers, who are legally entitled to kill feral cats and can always claim that they couldn't tell the difference. In one court case, it was accepted that although a suite of characteristics – tail stripes, tail length, body size – will help to identify a wildcat, these exist on a continuum so that distinguishing a pure wildcat from a hybrid or a feral cat is nigh-on impossible. You'd think that the precautionary principle would apply: if there's any chance that the animal's a wildcat, it shouldn't be killed. Instead, the precautionary principle seems to work the other way: if there's a chance that the animal *isn't* a wildcat, gamekeepers still seem able to kill it with impunity.

No sooner was the wildcat extinct than plans to reintroduce it got underway. After its success with pine martens, the Vincent Wildlife Trust is one of several groups now searching for suitable sites across England and Wales. They need remote areas with the right habitat, few roads and few humans – and therefore few domestic cats. There are several reasons why this project should prove even trickier than for pine martens. The first is the impossibility of sourcing sufficient quantities of wild-born wildcats from anywhere in Europe. Captive-bred animals won't have experience of hunting unless they've been raised in massive enclosures that naturally contain enough mice and voles to provide a complete diet; it's illegal to feed them live prey while they're in captivity. Along with all the usual stresses of being released, they'll suddenly need to catch their food, rather than having it handed to them, or tied to a string that a human half-heartedly jerks around an enclosure through the wire mesh. That's not the end of the problems. Even if they learn to hunt successfully, wildcats will need to be present in sustainable quantities from the start. As part of a healthy population, a wildcat is far more inclined to attack a feral cat than mate with it, but when numbers are low it's any port in a storm. These dangerous liaisons produce a hybrid population gravitating away from the forests and towards human settlements. In just a generation or two, the descendants of the fiercely proud wildcat will become skulkers and lurkers, frequenters of barns and outhouses.

*

When a lynx escaped from a Welsh animal park in 2017, the entire episode was depressingly instructive. Ceredigion Council decided that the risk to human life was 'severe', so it despatched a team of marksmen, one of whom shot the lynx while it was sleeping under a caravan. The marksman was later quoted as saying that 'action had to be taken'. We might call this the 'political passive', and file it alongside that hoary old classic: 'mistakes were made'. Quite why action was necessary has never been coherently explained; there was just the usual hysterical babble about safety being paramount. So let's compare lynx with dogs and cats. Every year in Britain, a quarter of a million people need medical treatment after a dog attack, and the small handful of fatalities are disproportionally likely to be infants inside their own homes. Lynx became extinct in Britain about 1,400 years ago, but they still live wild in Italy, Bulgaria, Sweden, Romania, Spain, Portugal and many other countries. There's not a single record of a wild lynx attacking humans – nowhere, ever, under any circumstance. A woman in the USA was reportedly given a head wound by a *pet* lynx in 2014. That's the sum total of lynx-on-human violence. A study in Bologna showed that 17.9 out of every 100,000 citizens will attend the emergency department in any given year with an injury inflicted by a domestic cat. By his own logic, that marksman would be better employed patrolling the city streets, taking potshots at our beloved pets wherever he encounters them. You can't be too careful, after all.

The cruel demise of Lillith the lynx lends credence to George Monbiot's claim that Britain may be 'the most zoophobic nation in Europe'. Greed makes an ally of fear. So in response to proposals that the species could be reintroduced to Britain, the then deputy president (now president) of the National Farmers' Union, Minette Batters, announced that lynx would create 'a danger to the public on walks around the countryside'. There are only two possibilities here: the NFU believes what it's saying because it's clueless, or it's indulging in brazenly Trumpist denials of reality in order to confuse the public and scare it into compliance. If so, it's emboldened to do so because it finds unfailing support from those pop-up politicians who make a career out of pandering to the most reactionary elements of the farming industry. Take the example of Fergus Ewing MSP, who, as Rural Economy Minister, told NFU Scotland delegates in 2018 that wolves, bears and lynx would be reintroduced 'over [his] dead body'. Presumably that wasn't meant as a challenge.

Unlike many of its more enlightened members, who are innovating in their agricultural methods and achieving extraordinary results for the benefit of biodiversity, the NFU remains trapped in a discredited philosophy that impoverishes the very landscapes it claims to protect. Public attitudes are shifting, but the same can't yet be said for the powerbrokers in the farming lobby. In 2021, another NFU representative baldly stated that the union opposed the reintroduction into Britain of any species. Can you imagine a more dismal policy than that, or a meaner vision for the future of our country? They make a desolation and call it peace.

It's Christmas in the Abruzzo National Park, and we're joined by a full complement of offspring: our daughters travelled with us, and our son has arrived direct from his student digs in the Netherlands. The weather is relatively mild, so the wildlife is active. Wolves are what we're here for. Oh, and bears. And chamois, and red deer, and wild boar, and stone martens, and golden eagles, and porcupines, and . . .

You often hear people excusing our country's abysmal record on biodiversity loss by claiming that it's unavoidable: Britain is simply too small and too densely populated. Cold hard facts sink that argument straight away: the Cairngorms, for example, is nearly ten times bigger than the Abruzzo National Park, despite being home to fewer people. The Italians aren't cowering under tables in case the wolves or the bears come after them. Just over an hour's drive from Rome, people are peacefully coexisting with apex predators, as they've done for generations. And well they might: vast sums of money are flooding into the area thanks to ecotourism.

There are probably ten or eleven wolf packs active in Abruzzo right now. Helped by our local guide, Valeria, we've found signs of them everywhere. Wolves, unlike dogs, run in a straight line, and the second wolf will often step all-but-precisely into the prints left by the leader. As there's snow on the ground, it's easy enough to follow their tracks; yesterday they led us to the bones of a roe deer that the wolves had killed the previous evening. There are bear prints and scats. There are chamois on the mountain ridges. There are herds of wild boar, sixty at a time, crossing hillsides in the distance. There are majestic red deer – including, up close the day before yesterday, a fine old stag with one antler completely gone.

Why can't we have this in Britain? The very idea of reintroducing wolves seems preposterous at the moment; we don't even seem able to keep hen harriers from being poisoned or blasted out of the sky. Yet if it weren't for the Channel, wolves would already have recolonised. They live across the continent in Portugal, Spain, France, Italy, Belgium, the Netherlands, Switzerland, Germany, Austria, all of Scandinavia, and further east towards (and including) Russia. Those economies haven't collapsed, and their children haven't been eaten. Meanwhile, the governments in Stormont, Cardiff and Westminster continue to behave like a man who calls himself a feminist in his Tinder bio. They know what we want to hear, but their much-vaunted credentials aren't borne out by discernible actions.

We haven't seen wolves or bears, but driving back from an early-morning trek today we spotted something even more improbable. Watching us by the side of the road was a wildcat. As we slowed to a halt, it stayed for a good ten seconds before vanishing down the bank and away. Valeria sounded as excited as we were. '*Il fantasma dei boschi*,' she whispered, unbelievingly. *Il fantasma dei boschi*: the ghost of the woods.

Beautiful, isn't it?

APPENDIX

Red List and Population Review

If you had to rank the species in most urgent need of intervention, where would you start? We believe intuitively that rare things must be more precarious than common ones, but rare compared with what? We'd expect lions to have smaller populations than the zebras they eat, and zebras will be rarer than most plants or insects. What about populations that are in steep decline, or highly fragmented, or limited to a tiny range? No single measure encapsulates all the different elements of risk. This gives scientists and governments the opportunity to cherry-pick or spin data according to their own agendas.

After wrangling with the problem for years, the International Union for Conservation of Nature (IUCN) came up with a standardised set of rules to be applied consistently across species. The result is known as the Red List. Each species – be it a dandelion, a white-bellied pangolin, or a monarch butterfly – is assessed against five criteria, and its conservation status is determined by the criterion with the lowest score. Cutting through 150 pages of detailed guidance on how to conduct the assessments, we can summarise the criteria under broad headings as follows:

A) Declining population
B) Small distribution *and* either decline or fluctuation
C) Small population size *and* decline
D) Very small or restricted population
E) Quantitative analysis (i.e. mathematical models to assess extinction risk)

Most Red Lists only consider species that have been present since at least the year 1500. This date approximates to the start of modern history, before which habitats and weather patterns may have differed. You won't find a British Red List assessment of lynx, because the last secure date for that species comes from bones carbon-dated to some time in the period 420–600 CE. Likewise, European brown bears had vanished from Britain by 594 CE. So these species can't even be classified as Extinct in the Wild, whereas the European wolf is included because it survived here until the eighteenth century.

Although the application of the 1500 cut-off date is fairly well accepted, there's more discussion over whether long-established but non-native species – those we term 'naturalised' – should be assessed at all. Among them are the rabbit and the brown hare, both of which have become an integral part of our British fauna. We've incorporated naturalised species into the population review, but our government agencies don't currently acknowledge their Red List status. The Netherlands takes a different approach, and welcomes these latecomers to its assessments.

The main purpose of the Red List is to galvanise conservation efforts by working out which species are at greatest threat of imminent extinction. Quite what 'imminent' means is never explicitly defined, but a working definition of twenty-five to thirty-five years is reasonable for mammals; the assessments themselves are largely based on changes over the preceding ten years or three generations (whichever happens to be longer). Red List assessments are made at a global, regional or national level, using the same processes regardless of geographical scale. For assessments at regional and national levels, we also consider whether incomers from another country or region could rescue flagging populations. Movement of species takes place regularly between our home nations, and Britain's island status doesn't quite exclude the possibility of migration even among land-based mammals: four Nathusius's pipistrelles were recently found to have travelled over 1,400 kilometres from Latvia to England.

Each of our native species is assigned to a category that runs along the scale of desperation from 'Extinct' to 'Least Concern', the intervening categories being (in order) 'Extinct in the Wild', 'Critically Endangered', 'Endangered', 'Vulnerable' and 'Near Threatened'. The collective badge 'Threatened' applies to everything that isn't classified as Near Threatened or Least Concern. Of the 8,431 British species assessed so far, fifteen percent are 'Threatened'. Unsurprisingly, 'Near Threatened' brings together species that slip just below the Threatened threshold, as well as those in danger of becoming Threatened very soon if we don't take action. Applying this system, we can agree that, for example, savanna elephants globally are Endangered owing to their declining populations (criterion A), and, by exactly the same measure, so are water voles in Britain.

There's one remaining category: 'Data Deficient'. Scientists love nothing better than to claim that they're lacking critical data, especially when they want more funding for research. Even so, the IUCN implores us to set aside our natural tendencies, and to refuse the temptation of the Data Deficient label if it's possible to make an assessment based on any single criterion. Experience shows that Data Deficient species tend to be ignored by politicians. Shrugging our shoulders and saying that we don't know what's happening is hardly an effective call to arms, whereas declaring that an animal is threatened with extinction earns us a prominent interview on the *Today* programme. In Britain we have four species officially considered Data Deficient. Three are closely related bat species (Alcathoe, Brandt's and whiskered) that even experts struggle to identify; without genetic testing, the only hope is to peer at the bumps on their teeth. The fourth is wild boar, and the squabble over their provenance in Britain means that, for official purposes, they're also classified as Data Deficient; the Mammal Society's assessment is Near Threatened.

The strategy of avoiding the Data Deficient category does have one unfortunate consequence. It's common for a species to be declining rapidly even though its geographical range remains unchanged. If there's no research to detect its population trend, it'll be dropped into the Least Concern category. That's why, in Britain at least, beetles are less commonly classified as Threatened than better-studied groups such as birds. Even popular species such as otters have remarkable data deficiencies. We know that since the 1970s they've gradually recolonised Britain, but it's not really possible to translate this into any sensible statement about total numbers. On a reliability scale of zero to five, where five indicates high confidence and zero no information at all, the otter population estimate scores 1. The new National Otter Survey of England that Fiona is currently coordinating will help a lot but won't provide definitive answers, because we don't know to what extent the identifiable signs of otters (footprints, spraint and so on) convert into numbers of animals.

According to IUCN criteria, rarity doesn't ensure a Threatened status unless there are fewer than 1,000 individuals left. What matters much more is whether the population is declining and, if so, how fast. This approach highlights to policymakers the species that need urgent attention, but it risks ignoring more gradual declines. After all, whatever piste you choose – black, red or blue – you're still heading to the bottom of the mountain. Another consequence is that shifting baselines aren't taken into account. In Britain, the pine marten suffered years of persecution that suppressed its populations to chronically low levels, but it's classified as 'Least Concern' because in recent years legislation has stopped the decline – even though its numbers are well below historical norms. Least Concern doesn't mean that we needn't feel concerned.

There are also species that, while suffering long and steady decline, aren't plummeting so catastrophically as to warrant inclusion on a Red List. These are the trickiest of all, because taking a long-term view is, by definition, never such an urgent priority as rescuing a species from imminent oblivion. Most of our small mammal species are gradually disappearing. Fiona's own research has shown that populations of bank voles, field voles, water shrews and common shrews are now declining at a rate of more than one percent per annum. A multitude of factors, including habitat degradation, a reduction in invertebrate prey owing to pesticide use, and the impacts of non-native species like rats and pheasants, are likely to play a part. But imperceptible declines with no single cause or quick fix attract very little publicity.

Red List status isn't just about population trends and absolute numbers. Range is also crucial. We know, for example, that hedgehogs and moles are found in almost every ten-by-ten-kilometre grid square of Britain, whereas the grey long-eared bat inhabits fewer than fifty grid squares, all of which are in southern England and (almost without exception) close to the coast. An extremely restricted or patchy distribution is risky because the species becomes vulnerable to inbreeding and chance events. Changes in distribution can reveal emergencies long before anyone gets round to a headcount. Thankfully, the story doesn't always have to be bleak: whereas red squirrels and water voles have been rapidly losing ground in England and Wales, the polecat has recovered much of its former range across the same landscapes.

It would take an awfully long time to count every hedgehog or polecat in Britain, even if we were able to find them all. The only solution is to estimate. For dormice and some bat species, volunteers carry out annual surveys of the animals at particular sites, and the information allows inferences about whether population sizes are changing. Working out absolute numbers, rather than just the trend, is an order of magnitude more difficult. For a lot of species, such as otters and shrews, frankly we have little or no idea. Our best guess for water shrews is about 700,000, but there could be as few as 240,000 or as many as 1.9 million. As for otters, we can't even place our estimate of 11,000 individuals within plausible ranges. Then there are – or maybe aren't – wild boar; suffice it to say that counting mammals can sometimes become *very* political, especially when the species in question isn't one that the government is altogether keen to protect.

Red Lists provide helpful warning systems, but governments and conservationists are well advised to use additional sources of information whenever possible. Every twenty years or so, the British government commissions a review of population status and trends for all our mammal species. The table that follows is a distillation of the results provided to the Statutory Nature Conservation Organisations by Fiona and her Mammal Society colleagues in 2018, along with the subsequent Red List status for each species.

Table showing population size (with reliability score), trends in population size and range, and Red List status for Britain's native and naturalised species.

'Naturalised' is defined as having arrived in Britain with the assistance of humans after the formation of the English Channel but before the twelfth century.

Information is shown only for the countries in which a given species occurs.

			Population size		Trends		
Group	Species	Country	Central estimate[1]	Reliability score[2]	Population size	Range	Red List[3]
Erinaceomorpha	Hedgehog	England	597,000	2.0	↓	=	VU
		Scotland	196,000				VU
		Wales	86,800				VU
		Britain	**879,000**				VU
Soricomorpha	Mole	England	24,300,000	1.0	=	=	LC
		Scotland	12,200,000				LC
		Wales	4,900,000				LC
		Britain	**41,400,000**				LC
	Common shrew	England	11,000,000	1.0	=	=	LC
		Scotland	7,690,000				LC
		Wales	2,330,000				LC
		Britain	**21,100,000**				LC
	Pygmy shrew	England	3,690,000	0.5	=	=	LC
		Scotland	1,430,000				LC
		Wales	1,170,000				LC
		Britain	**6,300,000**				LC
	Water shrew	England	458,000	0.0	=	↑	LC
		Scotland	118,000				LC
		Wales	137,000				LC
		Britain	**714,000**				LC
	Lesser white-toothed shrew (Naturalised or Native)	England	14,000	1.0	↑	=	NT
		Britain	**14,000**				NT
Lagomorpha	European rabbit (Naturalised)	England	21,300,000	1.0	↓	=	[LC]
		Scotland	11,800,000				[VU]
		Wales	2,910,000				[NT]
		Britain	**36,000,000**				[NT]
	Brown hare (Naturalised)	England	454,000	3.0	?	=	[LC]
		Scotland	88,000				[LC]
		Wales	37,000				[LC]
		Britain	**579,000**				[LC]
	Mountain hare[4]	England	2,500	2.0	?	↑	[VU]
		Scotland	132,000				NT
		Britain	**135,000**				NT
Rodentia	Red squirrel	England	38,900	2.5	↓	↓	EN
		Scotland	239,000				NT
		Wales	9,200				EN
		Britain	**287,000**				EN
	Beaver[5]	England	10	n/a	↑	↑	CR
		Scotland	158				EN
		Britain	**168**				EN
	Hazel dormouse	England	757,000	2.0	↓	=	VU
		Scotland	0				VU
		Wales	172,000				VU
		Britain	**930,000**				VU

			Population size		Trends		
Group	Species	Country	Central estimate[1]	Reliability score[2]	Population size	Range	Red List[3]
Rodentia	Bank vole	England	19,100,000	1.7	?	=	LC
		Scotland	5,390,000				LC
		Wales	2,930,000				LC
		Britain	**27,400,000**				LC
	Field vole	England	28,600,000	2.0	?	=	LC
		Scotland	21,500,000				LC
		Wales	9,760,000				LC
		Britain	**59,900,000**				LC
	Orkney vole	Scotland	?	0.0	↓	=	VU
		Britain	**?**				VU
	Water vole	England	77,000	3.0	↓	=	EN
		Scotland	50,000				NT
		Wales	4,500				EN
		Britain	**132,000**				EN
	Harvest mouse	England	532,000	0.0	?	?	LC
		Scotland	?				CR
		Wales	34,000				VU
		Britain	**566,000**				NT
	Wood mouse	England	22,700,000	2.0	=	=	LC
		Scotland	12,300,000				LC
		Wales	4,600,000				LC
		Britain	**39,600,000**				LC
	Yellow-necked mouse	England	1,360,000	2.5	?	↑	LC
		Scotland	0				LC
		Wales	140,000				LC
		Britain	**1,500,000**				LC
	House mouse (Naturalised)	England	4,340,000	2.0	=	=	[LC]
		Scotland	524,000				[LC]
		Wales	339,000				[LC]
		Britain	**5,203,000**				[LC]
	Black rat (Naturalised)	England	?	0.0	↓	↓	[CR]
		Scotland	?				[CR]
		Wales	?				[CR]
		Britain	**?**				[CR]
Carnivora	Wildcat[5]	England	0	2.0	↓	↓	RE
		Scotland	200				CR
		Wales	0				RE
		Britain	**200**				CR
	Red fox	England	255,000	2.5	?	=	LC
		Scotland	74,000				NT
		Wales	28,000				LC
		Britain	**357,000**				LC
	European wolf	England	0				RE
		Scotland	0				RE
		Britain	**0**				RE

			Population size		Trends		
Group	Species	Country	Central estimate[1]	Reliability score[2]	Population size	Range	Red List[3]
Carnivora	Badger	England	384,000	4.0	↑	=	LC
		Scotland	115,000				LC
		Wales	63,000				LC
		Britain	**562,000**				LC
	Otter	England	2,900	1.0	↑	↑	LC
		Scotland	7,100				VU
		Wales	1,000				VU
		Britain	**11,000**				LC
	Pine marten	England	?	3.0	↑	↑	CR
		Scotland	3,700				LC
		Wales	39				CR
		Britain	**3,700**				LC
	Stoat	England	260,000	1.0	?	=	LC
		Scotland	140,000				LC
		Wales	38,000				NT
		Britain	**438,000**				LC
	Weasel	England	308,000	?	?	=	LC
		Scotland	106,000				LC
		Wales	36,000				LC
		Britain	**450,000**				LC
	Polecat	England	66,000	4.0	↑	↑	LC
		Scotland	?				EN
		Wales	17,000				LC
		Britain	**83,000**				LC
Artiodactyla	Wild boar	England	500	2.0	↑	↑	[VU]
		Scotland	2,000				[VU]
		Wales	150				[EN]
		Britain	**2,600**				[NT]
	Red deer	England	80,000	4.0	↑	↑	LC
		Scotland	256,000				LC
		Wales	10,000				LC
		Britain	**346,000**				LC
	Fallow deer (Naturalised)	England	188,000	3.0	↑	↑	[LC]
		Scotland	57,000				[LC]
		Wales	19,000				[LC]
		Britain	**264,000**				[LC]
	Roe deer	England	120,000	4.0	↑	↑	LC
		Scotland	122,000				LC
		Wales	22,000				LC
		Britain	**265,000**				LC
Chiroptera	Greater horseshoe	England	10,000	4.0	↑	↑	LC
		Scotland	0				NA
		Wales	2,700				NT
		Britain	**13,000**				LC
	Lesser horseshoe	England	19,000	3.0	↑	↑	LC
		Scotland	0				LC
		Wales	31,000				LC
		Britain	**50,000**				LC

Group	Species	Country	Population size: Central estimate[1]	Population size: Reliability score[2]	Trends: Population size	Trends: Range	Red List[3]
Chiroptera	Alcathoe	**Britain**	**?**	0.0	?	?	DD
	Whiskered	**Britain**	**?**	0.0	?	?	DD
	Brandt's	**Britain**	**?**	0.0	?	?	DD
	Bechstein's	England	22,000	2.0	?	?	LC
		Scotland	0				NA
		Wales	250				EN
		Britain	**22,000**				LC
	Daubenton's	England	682,000	1.0	?	=	LC
		Scotland	235,000				LC
		Wales	108,000				LC
		Britain	**1,030,000**				LC
	Greater mouse-eared	England	1?	0.0	=	=	CR
		Britain	**1?**				CR
	Natterer's	England	321,000	2.0	?	?	LC
		Scotland	41,000				LC
		Wales	52,000				LC
		Britain	**414,000**				LC
	Serotine	England	117,000	3.0	?	?	VU
		Scotland	0				NA
		Wales	18,700				VU
		Britain	**136,000**				VU
	Leisler's	England	?	0.0	?	?	NT
		Scotland	?				NT
		Wales	?				NT
		Britain	**?**				NT
	Noctule	England	565,000	0.0	?	?	LC
		Scotland	42,000				LC
		Wales	92,000				LC
		Britain	**700,000**				LC
	Common pipistrelle	England	1,870,000	2.0	?	?	LC
		Scotland	875,000				LC
		Wales	297,000				LC
		Britain	**3,040,000**				LC
	Soprano pipistrelle	England	2,980,000	2.0	?	?	LC
		Scotland	1,210,000				LC
		Wales	478,000				LC
		Britain	**4,670,000**				LC
	Nathusius's pipistrelle	England	?	0.0	?	?	NT
		Scotland	?				VU
		Wales	?				VU
		Britain	**?**				NT
	Barbastelle	England	?	0.0	?	?	VU
		Wales	?				VU
		Britain	**?**				VU
	Brown long-eared	England	607,000	2.0	?	?	LC
		Scotland	230,000				LC
		Wales	97,000				LC
		Britain	**934,000**				LC
	Grey long-eared	England	1,000	1.0	?	↓	CR
		Britain	**1,000**				CR

1 Upper and lower plausible estimates are provided in the *Review of the Population and Conservation Status of British Mammals* (Mathews et al. 2018). Because of rounding, the totals may differ slightly from the sum of the estimates for country.
2 A score of 1 is low and 5 is high.
3 RE Regionally extinct; CR Critically Endangered; EN Endangered; VU Vulnerable; NT Near Threatened; LC Least Concern; DD Data Deficient. Classifications in brackets, although not adopted by British government agencies, are produced according to the same IUCN criteria. Wild boar are officially classified as DD owing to the debate about provenance.
4 Mountain hare is naturalised in England.
5 Since 2018, when the review was published, beaver populations in England and Scotland have increased substantially, and wildcat populations have very possibly disappeared altogether.

Acknowledgements

Much of the work underpinning this book was only possible thanks to Richard Shore. A wonderful biologist and friend, Richard passed away much too soon in 2019. We hope that, in some small way, we've been able to pay tribute to his legacy here.

Jim Gill and Amber Garvey of United Agents believed in us from the start. We've also been immensely fortunate with our editors Cecilia Stein and Rida Vaquas, who told us where we were going right and where we needed to change our ways: the book is much stronger for their advice. Jacqui Lewis copy-edited the typescript and saved us from countless embarrassing errors, alongside Tony Hirst, our excellent proofreader, and Laura McFarlane, who managed the process. Matilda Warner as press officer, and Lucy Cooper as marketing manager, helped us to reach our readers.

Various friends have read and commented on individual chapters. Many thanks to Dave and Sue Smallshire, John Gurnell, David Slater, Henry and Bridgit Schofield, Susan Burchett, Les Kendall, and Rob and Kate Gordon.

We're indebted to many other people who provided help or information, including Paul Chanin, Derek Gow, Kate Hills, Patrick Wright, Hugh Gillings, Johnny Birks, Merryl Gelling,

Lizzie Croose, Jenny MacPherson, Peter Ansell, Dean Jones, Robyn Stewart, Dave Garner, Cath Scott, Sandro Bertolino, Valeria Roselli, Tina and Tony Bricknell-Webb, Vicky Heaney, Robbie McDonald, Dave Garner, Domhnall Finch, Colin Harrower, Laura Kubasiewicz, Frazer Coomber, Mark Elliott, Derek Crawley, Alex Lees, Bethany Smith, Peter Evans, James Waggitt, Arabella Currie, Clive Flowers, Emyr Jenkins, and the members of the Wales Mammal Biodiversity Action Forum. Tony Mitchell-Jones and the mammal specialists of the Statutory Nature Conservation Organisations patiently shared the pain of endless revisions to the population review and the Red List. Thanks also to the good people of the Wiltshire Bat Group and the Mammal Society for your inspiration, insights, Werther's Originals and pastéis de nata.

We're grateful to the University of Sussex and the University of Exeter for keeping us in gainful employment while we wrote the book.

Our own pack of mammals – Charlie, Milly and Rosie – make only fleeting appearances in these chapters, but our adventures wouldn't have been anything like as much fun without them.

Notes

For a complete list of references, please visit https://britishmammals.com

PREFACE: I-SPY

1 Mathews, F., Kubasiewicz, L.M., Gurnell, J., Harrower, C.A., McDonald, R.A., and Shore, R.F., *A Review of the Population and Conservation Status of British Mammals. A report by the Mammal Society under contract to Natural England, Natural Resources Wales and Scottish Natural Heritage* (Peterborough: Natural England, 2018). The online version can be found at http://publications.naturalengland.org.uk/publication/5636785878597632.

2 Mathews, F. and Harrower, C., *IUCN-compliant Red List for Britain's Terrestrial Mammals. Assessment by the Mammal Society under contract to Natural England, Natural Resources Wales and Scottish Natural Heritage* (Peterborough: Natural England, 2020). The online version can be found at https://www.mammal.org.uk/science-research/red-list.

1. WHAT DID THE BEAVER SAY TO THE TREE?

1 Kosmider, R., Paterson, A., Voas, A., and Roberts, H., '*Echinococcus multilocularis* introduction and establishment in wildlife via imported beavers', *Veterinary Record*, 172, June 2013, p. 606.

2 Poliquin, R., *Beaver* (London: Reaktion Books, 2015), p. 88. For a full account of the trade in beaver fur, see pp. 81–122.

3 Alston, J.M., Maitland, B.M., Brito, B.T., Esmaeili, S., Ford, A.T., Hays, B., Jesmer, B.R., Molina, F.J., and Goheen, J.R., 'Reciprocity in restoration ecology: When might large carnivore reintroduction restore ecosystems?', *Biological Conservation*, 234, 2019, pp. 82–89.

4 Brazier, R.E., Elliott, M., Andison, E., Auster, R.E., Bridgewater, S., Burgess, P., Chant, J., Graham, H., Knott, E., Puttock, A.K., Sansum, P.,

and Vowles, A. (2020). *River Otter Beaver Trial: Science and Evidence Report* [Online], p. 70. https://www.wildlifetrusts.org/sites/default/files/2020-05/River%20Otter%20Beaver%20Trial%20-%20Science%20and%20Evidence%20Report.pdf. (Accessed 14 October 2021).

5 Williams, P., Biggs, J., Crowe, A., Murphy, J., Nicolet, P., Weatherby, A., and Dunbar, M. (2010). *Countryside Survey: Ponds Report from 2007* [Online]. https://nora.nerc.ac.uk/id/eprint/9622. (Accessed 1 September 2021).

6 See, for example, Puttock, A., Graham, H.A., Carless, D., and Brazier, R.E., 'Sediment and nutrient storage in a beaver engineered wetland', *Earth Surface Processes and Landforms*, 43, 2018, pp. 2358–2370.

7 VanSomeren, L., 'Scientists Are Relocating Nuisance Beavers to Help Salmon', *Smithsonian Magazine* [Online], 3 May 2021, https://www.smithsonianmag.com/science-nature/taking-nuisance-beavers-out-suburbs-can-help-save-salmon-180977491. (Accessed 4 November 2021).

8 Lohman, S., 'A Brief History of Castoreum, the Beaver Butt Secretion Used as Flavoring', *Mental Floss* [Online], 13 June 2017, https://www.mentalfloss.com/article/501813/brief-history-castoreum-beaver-butt-secretion-used-flavoring. (Accessed 30 October 2021).

9 Press Association, 'Fish farmer sues after otters "ate him out of house and home"', *Guardian* [Online], 26 February 2013, https://www.theguardian.com/environment/2013/feb/26/fish-farmer-environment-agency-otters. (Accessed 4 March 2021).

2. CRASHING BOARS

1 NatureScot (2022). *Updated non-native species risk assessment of feral pigs in Scotland.* [Online]. https://www.nature.scot/doc/naturescot-research-report-1288-updated-non-native-species-risk-assessment-feral-pigs-scotland. NatureScot Research Report, 1288. (Accessed 10 September 2022).

2 Frantz, A.C., Massei, G., and Burke, T., 'Genetic evidence for past hybridisation between domestic pigs and English wild boars', *Conservation Genetics*, 13, October 2012, pp. 1355–1364.

3 IUCN (2012). *Guidelines for Reintroductions and Other Conservation Translocations.* [Online]. https://portals.iucn.org/library/efiles/documents/2013-009.pdf. The Reintroduction and Invasive Species Specialist Groups' Task Force on Moving Plants and Animals for Conservation Purposes, p. 6.

4 Vetter, S.G., Ruf, T., Bieber, C., and Arnold, W. (2015). 'What is a mild winter? Regional differences in within-species responses to climate change'.

[Online]. doi.org/10.1371/journal.pone.0132178. *PLOS One*, 10 (7). (Accessed 20 June 2022).

5 Vetter, S.G., Puskas, Z., Bieber, C., and Ruf, T., 'How climate change and wildlife management affect population structure in wild boars', *Scientific Reports*, 10 (1), 2020, pp. 1–10.

6 Lutwyche, Richard, *The Pig: A Natural History* (Princeton: Princeton University Press, 2020), p. 5.

7 See Hautzinger, D., 'Helen Macdonald on the Wonderful, Optimistic World of Hawks'. [Online]. https://interactive.wttw.com/playlist/2017/11/01/helen-macdonald-hawk. *WTTW*, 1 November 2017. (Accessed 16 August 2022).

8 Bateman, I.J., Day, B.H., *et al.* (2014). [Online]. *UK National Ecosystem Assessment Follow-on. Work Package Report 3: Economic value of ecosystem services.* UNEP-WCMC. (Accessed 2 October 2021).

9 Forestry England (2022). *Background, population and management of boar in the Forest of Dean.* [Online]. https://www.forestryengland.uk/article/more-information-about-wild-boar. (Accessed 16 August 2022). The original quotation has been slightly revised.

10 Stannard, K.G. (2011). *Feral Wild Boar Management Plan.* [Online]. https://img1.wsimg.com/blobby/go/52f17869-2a6b-4a99-abe4-1d4d3057ecc3/downloads/1cv5vq595_934382.pdf?ver=1600874984081. Forestry Commission. (Accessed 19 January 2022).

11 Monbiot, G., *Feral: Rewilding the Land, Sea and Human Life* (London: Allen Lane, 2013), p. 95.

12 Ballari, S.A. and Barrios-García, M.N., 'A review of wild boar *Sus scrofa* diet and factors affecting food selection in native and introduced ranges', *Mammal Review*, 44 (2), October 2013, pp. 124–134.

13 Rebanks, J., Public lecture at the University of Exeter, 15 November 2021.

14 Sandom, C.J., Hughes, J., and Macdonald, D.W., 'Rooting for rewilding: quantifying wild boar's *Sus scrofa* rooting rate in the Scottish Highlands', *Restoration Ecology*, 21 (3), 2013, pp. 329–335.

15 Forestry England (2022). *Background, population and management of boar in the Forest of Dean.* [Online]. https://www.forestryengland.uk/article/more-information-about-wild-boar. (Accessed 16 August 2022).

16 de Schaetzen, F., van Langevelde, F., and WallisDeVries, M.F., 'The influence of wild boar (*Sus scrofa*) on microhabitat quality for the endangered butterfly *Pyrgus malvae* in the Netherlands', *Journal of Insect Conservation*, 22, 2018, pp. 51–59.

17 TB hub, *TB in wildlife.* [Online]. https://tbhub.co.uk/tb-in-wildlife. (Accessed 16 August 2022).

18 Department for Environment, Food and Rural Affairs, Scottish Government, and Welsh Government. (August 2014, updated March 2020). *Disease control*

strategy for African and Classical Swine Fever in Great Britain. [Online]. https://assets.publishing.service.gov.uk/government/uploads/system/uploads/attachment_data/file/877081/disease-control-strategy-csf-2020a.pdf. (Accessed 15 August 2022).

19 Van Dooren, K. (29 July 2020). 'ASF is a growing problem for Poland'. [Online]. *Pig Progress.* https://www.pigprogress.net/health-nutrition/asf-is-a-growing-problem-for-poland. (Accessed 16 August 2022).

20 O'Mahony, K. (2019). *Feral Bo(a)rderlands: living with and governing wild boar in the Forest of Dean.* [Online]. https://orca.cardiff.ac.uk/131010/1/Feral%20Bo%28a%29rderlands-%20living%20with%20and%20governing%20wild%20boar%20in%20the%20Forest%20of%20Dean.pdf. PhD thesis, University of Cardiff. (Accessed 24 February 2021).

3. ON THE TRAIL OF THE LONESOME PINE MARTEN

1 Birks, J., *Pine Martens* (Epping: Whittet Books, 2017), p. 7.

2 Ibid., pp. 22–25.

3 Lindenfors, P., Dalèn, L., and Angerbjörn, A., 'The monophyletic origin of delayed implantation in carnivores and its implications', *Evolution*, 57 (8), August 2003, pp. 1952–1956.

4 Birks, *Pine Martens*, p. 5.

5 Fischer, J. and Lindenmayer, D.B., 'An assessment of the published results of animal relocations', *Biological Conservation*, 96 (1), November 2000, pp. 1–11; Wolf, C.M., Garland Jr., T., Griffith, B., 'Predictors of avian and mammalian translocation success: reanalysis with phylogenetically independent contrasts', *Biological Conservation*, 86 (2), November 1988, pp. 243–255.

6 Pullar, P., *A Richness of Martens: Wildlife Tales from the Highlands* (Edinburgh: Birlinn, 2020), p. 41.

7 MacPherson, J., *Feasibility Assessment for Reinforcing Pine Marten Numbers in England and Wales* (Ledbury: Vincent Wildlife Trust, 2014), pp. 51–58; Bavin, D., MacPherson, J., Denman, H., Crowley, S.L., and McDonald, R.A., 'Using Q-methodology to understand stakeholder perspectives on a carnivore translocation', *People and Nature*, 2 (4), 2020, pp. 1117–1130.

8 Sheehy, E., Sutherland, C., O'Reilly, C., and Lambin, X. (2018). 'The enemy of my enemy is my friend: native pine marten recovery reverses the decline of the red squirrel by suppressing grey squirrel populations'. [Online]. doi.org/10.1098/rspb.2017.2603. *Proceedings of the Royal Society B: Biological Sciences*, 285. (Accessed 23 March 2022).

9 Twining, J.P., Montgomery, W.I., Price, L., Kunc, H.P., and Tosh, D.G. (2020). 'Native and invasive squirrels show different behavioural responses to scent of a shared native predator'. [Online]. doi/10.1098/rsos.191841. *Royal Society Open Science*, 7. (Accessed 25 March 2022).

10 Chow, P.K., Clayton, N.S., and Steele, M.A. (2021). 'Cognitive performance of wild eastern gray squirrels (*Sciurus carolinensis*) in rural and urban, native, and non-native environments'. [Online]. doi.org/10.3389/fevo.2021.615899. *Frontiers in Ecology and Evolution*, 9. (Accessed 18 March 2022).

11 Twining, J.P., Montgomery, W.I., and Tosh, D.G., 'The dynamics of pine marten predation on red and grey squirrels', *Mammalian Biology*, 100, 2020, pp. 285–293.

12 McNicol, C.M., Bavin, D., Bearhop, S., Ferryman, M., Gill, R., Goodwin, C.E., MacPherson, J., Silk, M.J., and McDonald, R.A., 'Translocated native pine martens *Martes martes* alter short-term space use by invasive non-native grey squirrels *Sciurus carolinensis*', *Journal of Applied Ecology*, 57, 2020, pp. 903–913.

13 Ciuti, S., Northrup, J.M., Muhly, T.B., Simi, S., Musiani, M., Pitt, J.A., Boyce, M.S. (2012). 'Effects of humans on behaviour of wildlife exceed those of natural predators in a landscape of fear'. [Online]. doi.org/10.1371%2Fjournal.pone.0050611. *PLOS One*, 7(11). (Accessed 1 June 2022).

14 McNicol *et al.*, 'Translocated native pine martens *Martes martes* alter short-term space use by invasive non-native grey squirrels *Sciurus carolinensis*', pp. 903–913.

15 Baines, D., Moss, R., and Dugan, D., 'Capercaillie breeding success in relation to forest habitat and predator abundance', *Journal of Applied Ecology*, 41, 2004, pp. 59–71.

4. WATER VOLES AND EARTH HOUNDS

1 Melero, Y., Robinson, E., and Lambin, X., 'Density-and age-dependent reproduction partially compensates culling efforts of invasive non-native American mink', *Biological Invasions*, 17 (9), 2015, pp. 2645–2657.

2 Strachan, C., Strachan, R., and Jeffries, D.J., *Preliminary report on the changes in the water vole population of Britain as shown by the national surveys of 1989–1990 and 1996–1998* (Ledbury: Vincent Wildlife Trust, 2000).

3 Dean, M., Strachan, R., Gow, D., and Andrews, R., *The Water Vole Mitigation Handbook*, ed. Mathews, F. and Chanin, P. (London: Mammal Society, 2016).

4 Brzeziński, M., Chibowska, P., Zalewski, A., Borowik, T., and Komar, E., 'Water vole *Arvicola amphibius* population under the impact of the American

mink *Neovison vison*: Are small midfield ponds safe refuges against this invasive predator?', *Mammalian Biology*, 93 (1), 2018, pp. 182–188.

5 Moorhouse, T.P., Gelling, M., and Macdonald, D.W., 'Effects of Forage Availability on Growth and Maturation Rates in Water Voles', *Journal of Animal Ecology*, 77 (6), 2008, pp. 1288–1295.

6 Moorhouse, T.P., Gelling, M., and Macdonald, D.W., 'Effects of habitat quality upon reintroduction success in water voles: evidence from a replicated experiment', *Biological Conservation*, 142 (1), 2009, pp. 53–60.

7 Stewart, R.A., Clark, T.J., Shelton, J., Stringfellow, M., Scott, C., White, S.A., and McCafferty, D.J. (2017). 'Urban grasslands support threatened water voles'. [Online]. https://academic.oup.com/jue/article/3/1/jux007/4097929. *Journal of Urban Ecology*, 3 (1). (Accessed 13 January 2021).

8 Brace, S., Ruddy, M., Miller, R., Schreve, D.C., Stewart, J.R., and Barnes, I. (2016). 'The colonization history of British water vole (*Arvicola amphibius* (Linnaeus, 1758)): origins and development of the Celtic fringe'. [Online]. doi.org/10.1098/rspb.2016.0130. *Proceedings of the Royal Society B: Biological Sciences*, 283. (Accessed 24 January 2021).

9 Strachan, R., *Water Voles* (Epping: Whittet Books, 1997), p. 58.

10 Kryštufek, B., Koren, T., Engelberger, S., Horváth, G., Purger, J., Arslan, A., Chişamera, G., and Murariu, D., 'Fossorial morphotype does not make a species in water voles', *Mammalia*, 79, 2015, pp. 293–303.

5. HANGING OUT WITH GREATER HORSESHOE BATS

1 Finch, D., Schofield, H., and Mathews, F., 'Habitat associations of bats in an agricultural landscape: Linear features versus open habitats', *Animals*, 10, 2020, pp. 1856–1866.

2 He, W., Pedersen, S.C., Gupta, A.K., Simmons, J.A., and Müller, R. (2015). 'Lancet Dynamics in Greater Horseshoe Bats, *Rhinolophus ferrumequinum*'. [Online]. doi.org/10.1371/journal.pone.0121700. *PLOS One*, 10 (4). (Accessed 12 June 2022).

3 Ma, J., Kobayasi, K., Zhang, S., and Metzner, W., 'Vocal communication in adult greater horseshoe bats, *Rhinolophus ferrumequinum*', *Journal of Comparative Physiology A.*, 192 (5), 2006, pp. 535–550.

4 Liu, Y., Metzner, W., and Feng, J., 'Vocalization during copulation behavior in greater horseshoe bats, *Rhinolophus ferrumequinum*', *Chinese Science Bulletin*, 58 (18), 2013, pp. 2179–2184.

5 Finch, D., Schofield, H., Firth, J.A., and Mathews, F., 'Social networks of the greater horseshoe bat during the hibernation season: a landscape-scale case study', *Animal Behaviour*, 188, 2022, pp. 25–34.

6 Coomber, F.G., Smith, B.R., August, T.A., Harrower, C.A., Powney, G.D., and Mathews, F. (2021). 'Using biological records to infer long-term occupancy trends of mammals in the UK'. [Online]. 10.1016/j.biocon.2021.109362. *Biological Conservation*, 264 (3). (Accessed 3 June 2022).

7 Park, K.J., Jones, G., and Ransome, R.D., 'Torpor, arousal and activity of hibernating greater horseshoe bats (*Rhinolophus ferrumequinum*)', *Functional Ecology*, 14, 2000, pp. 580–588.

8 Russo, D., Maglio, G., Rainho, A., Meyer, C.F., and Palmeirim, J.M., 'Out of the dark: diurnal activity in the bat *Hipposideros ruber* on São Tomé island (West Africa)', *Mammalian Biology*, 76 (6), 2011, pp. 701–708.

9 Finch, D., Corbacho, D.P., Schofield, H., Davison, S., Wright, P.G., Broughton, R.K., and Mathews, F., 'Modelling the functional connectivity of landscapes for greater horseshoe bats *Rhinolophus ferrumequinum* at a local scale', *Landscape Ecology*, 35, 2020, pp. 577–589.

10 Tournayre, O., Pons, J.B., Leuchtmann, M., Leblois, R., Piry, S., Filippi-Codaccioni, O., Loiseau, A., Duhayer, J., Garin, I., Mathews, F., and Puechmaille, S., 'Integrating population genetics to define conservation units from the core to the edge of *Rhinolophus ferrumequinum* western range', *Ecology and Evolution*, 9, 2019, pp. 12272–12290.

11 Rossiter, S.J., Jones, G., Ransome, R.D., and Barratt, E.M., 'Outbreeding increases offspring survival in wild greater horseshoe bats (*Rhinolophus ferrumequinum*)', *Proceedings of the Royal Society of London. Series B: Biological Sciences*, 268 (1471), 2008, pp. 1055–1061.

12 Pfeiffer, B. and Mayer, F., 'Spermatogenesis, sperm storage and reproductive timing in bats', *Journal of Zoology*, 289 (2), 2013, pp. 77–85.

13 Matsumura, S., 'Mother–infant communication in a horseshoe bat (*Rhinolophus ferrumequinum nippon)*: vocal communication in three-week-old infants', *Journal of Mammalogy*, 62 (1), 1981, pp. 20–28.

14 Wright, P.G., Kitching, T., Hanniffy, R., Palacios, M.B., McAney, K., and Schofield, H., 'Effect of roost management on populations trends of *Rhinolophus hipposideros* and *Rhinolophus ferrumequinum* in Britain and Ireland', *Conservation Evidence*, 19, 2022, pp. 21–26.

6. TIGGYWINKLE GOES ROGUE

1 Roos, S., Johnston, A., and Noble, D., 'UK hedgehog datasets and their potential for long-term monitoring', *BTO Research Report*, 598, 2012, pp. 1–63.

2 Wembridge, D., Newman, M.R., Bright, P., and Morris, P., 'An estimate of the annual number of hedgehog (*Erinaceus europaeus*) road casualties in Great Britain', *Mammal Communications*, 2, 2016, pp. 8–14.
3 Wright, P.G., Coomber, F.G., Bellamy, C.C., Perkins, S.E., and Mathews, F. (2020). 'Predicting hedgehog mortality risks on British roads using habitat suitability modelling'. [Online]. doi.org/10.7717/peerj.8154. *PeerJ*, 7. (Accessed 31 May 2021).
4 Doncaster, C.P., 'Testing the role of intraguild predation in regulating hedgehog populations', *Proceedings of the Royal Society of London. Series B: Biological Sciences*, 249 (1324), 1992, pp. 113–117.
5 Trewby, I.D., Young, R., McDonald, R.A., Wilson, G.J., Davison, J., Walker, N., Robertson, A., Doncaster, C.P., and Delahay, R.J. (2014). 'Impacts of removing badgers on localised counts of hedgehogs'. [Online]. doi.org/10.1371/journal.pone.0095477. *PLOS One*, 9 (4). (Accessed 12 March 2021).
6 Finch, D., Smith, B.R., Marshall, C., Coomber, F.G., Kubasiewicz, L.M., Anderson, M., Wright, P.G., and Mathews, F. (2020). 'Effects of Artificial Light at Night (ALAN) on European hedgehog activity at supplementary feeding stations'. [Online]. 10.3390/ani10050768. *Animals*, 10 (5). (Accessed 19 January 2021).
7 Gazzard, A. and Baker, P.J. (2020). 'Patterns of feeding by householders affect activity of hedgehogs (*Erinaceus europaeus*) during the hibernation period'. doi.org/10.3390/ani10081344. *Animals*, 10 (8). (Accessed 11 November 2021).
8 Keymer, I.F., Gibson, E.A., and Reynolds, D.J., 'Zoonoses and other findings in hedgehogs (*Erinaceus europaeus*): a survey of mortality and review of the literature', *The Veterinary Record*, 128 (11), 1991, pp. 245–249.
9 Dowding, C.V., Shore, R.F., Worgan, A., Baker, P.J., and Harris, S., 'Accumulation of anticoagulant rodenticides in a non-target insectivore, the European hedgehog (*Erinaceus europaeus*)', *Environmental Pollution*, 158 (1), 2010, pp. 161–166.
10 Reeve, N., *Hedgehogs* (London: T & A.D. Poyser, 1994), p. 167.
11 Warwick, H., *A Prickly Affair: The Charm of the Hedgehog* (London: Penguin, 2010), pp. 61–117. First published 2008.

7. WHO CARES WHAT COLOUR THE SQUIRRELS ARE?

1 Zenni, R.D., Essl, F., García-Berthou, E., and McDermott, S.M., 'The economic costs of biological invasions around the world', *NeoBiota*, 67, 2021, pp. 1–9.

2 Chantrey, J., Dale, TD., Read, J.M., White, S., Whitfield, F., Jones, D., McInnes, C.J., and Begon, M., 'European red squirrel population dynamics driven by squirrelpox at a gray squirrel invasion interface', *Ecology and Evolution*, 4 (19), October 2014, pp. 3788–3799.

3 Gurnell, J., 'Squirrel numbers and the abundance of tree seeds', *Mammal Review*, 13 (2-4), 1983, pp. 133–148.

4 Chantrey, J., Dale, T., Jones, D., Begon, M., and Fenton, A. (2019). 'The drivers of squirrelpox virus dynamics in its grey squirrel reservoir host'. [Online]. https://www.sciencedirect.com/science/article/pii/S1755436519300362. *Epidemics*, 28. (Accessed 10 October 2021).

5 Bright, P.W., 'Habitat fragmentation – problems and predictions for British Mammals', *Mammal Review*, 23, 1993, pp. 101–111.

6 Everest, D.J., Shuttleworth, C.M., Stidworthy, M.F., Grierson, S.S., Duff, P.J., and Kenward, R.E., 'Adenovirus: An emerging factor in red squirrel *Sciurus vulgaris* conservation', *Mammal Review*, 44, 2014, pp. 225–233.

7 Nichols, C.P., Drewe, J.A., Gill, R., Goode, N., and Gregory, N., 'A novel causal mechanism for grey squirrel bark stripping: the Calcium Hypothesis', *Forest Ecology and Management*, 367, 2016, pp. 12–20.

8 Bertolino, S., Currado, I., Mazzoglio, P., and Amori, G., 'Native and alien squirrels in Italy', *Hystrix: the Italian Journal of Mammalogy*, 11 (2), 2000, pp. 65–74.

9 Romeo, C., McInnes, C.J., Dale, T.D., Shuttleworth, C., Bertolino, S., Wauters, L.A., and Ferrari, N., 'Disease, invasions and conservation: no evidence of squirrelpox virus in grey squirrels introduced to Italy', *Animal Conservation*, 22 (1), 2019, pp. 14–23.

10 Rushton, S.P., Lurz, P.W., Gurnell, J., Nettleton, P., Bruemmer, C., Shirley, M.D., and Sainsbury, A., 'Disease threats posed by alien species: the role of a poxvirus in the decline of the native red squirrel in Britain', *Epidemiology & Infection*, 134 (3), 2006, pp. 521–533.

11 Kenward, R.E. and Holm, J.L., 'On the replacement of the red squirrel in Britain: a phytotoxic explanation', *Proceedings of the Royal Society B*, 251, 1993, pp. 187–194.

12 Wauters, L.A., Lurz, P.W.W., and Gurnell, J., 'The interspecific effects of grey squirrels (*Sciurus carolinensis*) on the space use and population demography of red squirrels (*S. vulgaris*) in conifer plantations', *Ecology Research*, 15, 2000, pp. 271–284.

13 Wauters, L.A., Tosi, G., and Gurnell, J., 'Interspecific competition in tree squirrels: do introduced grey squirrels (*Sciurus carolinensis*) deplete tree seeds hoarded by red squirrels (*S. vulgaris*)?', *Behaviour, Evolution and Sociobiology*, 51, 2002, pp. 360–367.

14 Wauters, L.A., Gurnell, J., Currado, I., and Mazzoglio, P.J., 'Grey squirrel management in Italy – squirrel distribution in a highly fragmented landscape', *Wildlife Biology*, 3, 1997, pp. 117–124.

15 Bertolino, S. and Genovesi, P., 'Spread and attempted eradication of the grey squirrel (*Sciurus carolinensis*) in Italy, and consequences for the red squirrel (*Sciurus vulgaris*) in Eurasia', *Biological Conservation*, 109 (3), 2003, pp. 351–358.
16 Wauters, L. and Martinoli, A., 'A golden cage for the European red squirrel in Italy? proposal for a targeted control of the grey squirrel', *Biodiversity*, 22 (1–2), 2021, pp. 87–90.
17 Scapin, P., Ulbano, M., Ruggiero, C., Balduzzi, A., Marsan, A., Ferrari, N., and Bertolino, S., 'Surgical sterilization of male and female grey squirrels (*Sciurus carolinensis*) of an urban population introduced in Italy', *The Journal of Veterinary Medical Science*, 81 (4), 2019, pp. 641–645.
18 Dunn, M., Marzano, M., and Forster, J. (2021). 'The red zone: Attitudes towards squirrels and their management where it matters most'. [Online]. doi.org/10.1016/j.biocon.2020.108869. *Biological Conservation*, 253.
19 Massei, G., Cowan, D., Eckery, D., Mauldin, R., Gomm, M., Rochaix, P., Hill, F., Pinkham, R., and Miller, L.A. (2020). 'Effect of vaccination with a novel GnRH-based immunocontraceptive on immune responses and fertility in rats'. [Online]. doi.org/10.1016/j.heliyon.2020.e03781. *Heliyon*, 6 (4).
20 Croft, S., Aegerter, J.N., Beatham, S., Coats, J., and Massei, G. (2021). 'A spatially-explicit population model to compare culling versus fertility control to reduce numbers of grey squirrels'. [Online]. doi.org/10.1016/j.ecolmodel.2020.109386. *Ecological Modelling*, 440.

8. A PHOCINE GOOD STORY: SAVING GREY SEALS

1 Allen, R., Jarvis, D., Sayer, S., and Mills, C., 'Entanglement of grey seals *Halichoerus grypus* at a haul out site in Cornwall, UK', *Marine Pollution Bulletin*, 64 (12), 2012, pp. 2815–2819.
2 Natural Environment Research Council Special Committee on Seals (SCOS). (2022). *Scientific advice on matters related to the management of seal populations: 2021*. [Online]. http://www.smru.st-andrews.ac.uk/files/2022/08/SCOS-2021.pdf. (Accessed 16 July 2022).
3 Sayer, S., Allen, R., Hawkes, L.A., Hockley, K., Jarvis, D., and Witt, M.J., 'Pinnipeds, people and photo identification: the implications of grey seal movements for effective management of the species', *Journal of the Marine Biological Association of the United Kingdom*, 99, 2019, pp. 1221–1230.
4 Zheng, X., Kamat, A.M., Cao, M., and Kottapalli, A.G.P. (2021). 'Creating underwater vision through wavy whiskers: a review of the flow-sensing mechanisms and biomimetic potential of seal whiskers'. [Online]. doi.

org/10.1098/rsif.2021.0629. *Journal of the Royal Society Interface*, 18. (Accessed 19 September 2022).

5 Willer, D.F., Robinson, J.P., Patterson, G.T., and Luyckx, K. (2022). 'Maximising sustainable nutrient production from coupled fisheries-aquaculture systems'. [Online]. doi.org/10.1371/journal.pstr.0000005. *PLOS Sustainability and Transformation*, 1 (3). (Accessed 12 September 2022).

6 McGinnity, P., Prodöhl, P., Ferguson, A., Hynes, R., Maoiléidigh, N.O., Baker, N., and Cross, T., 'Fitness reduction and potential extinction of wild populations of Atlantic salmon, *Salmo salar*, as a result of interactions with escaped farm salmon', *Proceedings of the Royal Society B: Biological Sciences*, 270 (1532), 2003, pp. 2443–2450.

7 Stansbury, A.L., Götz, T., Deecke, V.B., and Janik, V.M. (2015). 'Grey seals use anthropogenic signals from acoustic tags to locate fish: evidence from a simulated foraging task'. [Online]. doi.org/10.1098/rspb.2014.1595. *Proceedings of the Royal Society B: Biological Sciences*, 282. (Accessed 14 September 2022).

8 Thomas, L., Russell, D.J.F., Duck, C.D., Morris, C.D., Longergan, M., Empacer, F., Thompson, D., and Harwood, J., 'Modelling the population size and dynamics of the British grey seal', *Aquatic Conservation: Marine and Freshwater Ecosystems*, 29 (S1), 2019, pp. 6–23.

9 Thompson, D., Duck, C.D., Morris, C.D., and Russell, D.J.F., 'The status of harbour seals (*Phoca vitulina*) in the UK', *Aquatic Conservation: Marine and Freshwater Ecosystems*, 29 (S1), 2019, pp. 40–60.

10 Bolt, H.E., Harvey, P.V., Mandleberg, L., and Foote, A.D., 'Occurrence of killer whales in Scottish inshore waters: temporal and spatial patterns relative to the distribution of declining harbour seal populations', *Aquatic Conservation: Marine and Freshwater Ecosystems*, 19, 2009, pp. 671–675.

11 Sparling, C.E., Fedak, M.A., and Thompson, D., 'Eat now, pay later? Evidence of deferred food-processing costs in diving seals', *Biology Letters*, 3 (1), 2007, pp. 94–98.

12 Russell, D.J., McClintock, B.T., Matthiopoulos, J., Thompson, P.M., Thompson, D., Hammond, P.S., Jones, E.L., MacKenzie, M.L., Moss, S., and McConnell, B.J., 'Intrinsic and extrinsic drivers of activity budgets in sympatric grey and harbour seals', *Oikos*, 124 (11), 2015, pp. 1462–1472.

13 Onoufriou, J., Russell, D.J., Thompson, D., Moss, S.E., and Hastie, G.D., 'Quantifying the effects of tidal turbine array operations on the distribution of marine mammals: Implications for collision risk', *Renewable Energy*, 180, 2021, pp. 157–165.

14 Davies, B.F., Holmes, L., Rees, A., Attrill, M.J., Cartwright, A.Y., and Sheehan, E.V., 'Ecosystem Approach to Fisheries Management works – how switching from mobile to static fishing gear improves populations of

fished and non-fished species inside a marine-protected area', *Journal of Applied Ecology*, 58, 2021, pp. 2463–2478.

15 Lambert, R.A., 'The Grey Seal in Britain: a twentieth century history of a nature conservation success', *Environment and History*, 8 (4), 2002, pp. 449–474.

Index

acoustic deterrent devices (ADDs) 295
Aesop 23
afforestation 63–4, 78, 240–1
African swine fever (ASF) 68–9, 71, 72–4, 93
agriculture 12–13, 135, 176–7
  and hedgehogs 204, 205
  and pigs 69, 71
  and sheep 95
  and trees 240, 241
  and water voles 122
  and wild boar 44, 45, 60
  see also National Farmers' Union
Angling Trust 15–16, 17
Archilochus 217
aristocracy 56–7
Aristotle 216, 218, 275, 302
Aubrey, John 57
Australia 104
Austria 160

badgers 19–20, 63, 83, 87, 199–201
  and culling 68, 73, 311
bandicoots 104
Barcelona (Spain) 49, 72
bats 113, 156–8, 159–60, 183–4
  and greater mouse-eared ix, 79, 158–9, 314
  and Red List 324, 326–7
  see also greater horseshoe bats
Bats in Churches project 191
Batters, Minette 317
Bavaria (Germany) 6, 8, 28
beavers xiii, 1–5, 19–24, 28–33, 35–7
  and castoreum 24–5
  and disease 5–7
  and fish stocks 15–17, 18–19
  and fur trade 8–10
  and reintroduction 7–8, 33–5, 52, 92, 310
  and Scotland 25–7
  and Wales 27–8
  and wetlands 10–12, 13–15
Belgium 71, 73, 160
Berlin, Isaiah 217–18
biodiversity 54–5, 66–7, 206, 235–6, 318–19
birds 53–4, 223–5, 227, 285–6, 287
Birks, Johnny 84, 85, 89, 99
birth control 267–70
bluebells 65
boar, *see* wild boar
Boece, Hector 22–3
Boscastle (Cornwall) 189–92
bovine tuberculosis (TB) 68, 98, 200
Bradley, Richard 59
breeding:
  and bats 186–8
  and beavers 20, 26, 28, 31, 33
  and hedgehogs 217, 223
  and pine martens 87
  and red squirrels 246
  and seals 275, 277, 280–1, 286–7, 302, 306
  and water voles 131
  and wild boar 39, 41–2, 45, 46–7, 73
  and wildcats 314
  see also captive breeding
Brewer, William Jones 174
Britain, *see* England; Northern Ireland; Scotland; Wales
Brocklehurst, Thomas 236–7

Bulgaria ix, 6, 316
butterflies 66–7, 285

Canada 5. 9, 15, 84, 106–7, 276
capercaillies 108–12
captive breeding 253–4, 314, 315
castoreum 24–5
cats, *see* wildcats
cave paintings 43–4
Charles II of England, King 57
China 71
Chinese medicine 257
chromosomes 4
churches 190–1
Cicero 23
Clare, John 204
climate change 13–14, 46–7, 186–7
Coleridge, Samuel Taylor 1
Colvile, Oliver 193, 194
Coly River (Devon) 12, 13
Conservation of Seals Act (1970) 291
Covid-19 pandemic 5, 53, 81, 119–20, 126, 248
Croatia 54–5, 86–7, 241
Cromwell, Thomas 23
Crouch, Tracey 308
culling 46, 69, 70–1
  and badgers 68, 73, 200–1
  and grey squirrels 250, 260–3
  and hedgehogs 224–6
  and mink 126, 129, 135
  and pine martens 110, 111–12
  and red squirrels 242–4
  and seals 282, 291, 292–3, 296, 311

Darwin, Charles 48, 142
  The Origin of Species 274
Data Deficient status 40, 324–5
deer 105, 107, 228, 318–19
Defra 3–4, 7, 210–11, 269
  and wild boar 46, 58, 68, 70, 74–5
Denman, Huw 95
Denmark 73, 126
Devil's Bridge (Ceredigion) 80–2, 94, 114–17
Devon Wildlife Trust 3
disease 5–7, 93, 98, 137, 254
  and wild boar 59, 61–2, 68–71, 72–4
  see also squirrelpox
dogs 6, 7, 19, 50, 221, 316
dolphins 163, 232, 273, 305, 308
Dorset Mammal Group 199
Drayton, Michael: *Poly-Olbion* 22

'earth hounds' 124–5
East Anglia 130–1, 135
Echinococcus multilocularis (tapeworm) 6–7
echolocation 156, 157, 163–5, 302–3
ecosystems xiii, 10, 236, 241
  and marine 294, 297, 299
  and wild boar 62–6
Edward II of England, King 41
Edward III of England, King 256
elephants xii, 34, 40, 86, 272, 323
Endangered species xii, 25, 78, 108, 158–9
  and wild boar 39, 40
endothermy 175
E*NET*WILD 61–2
England 23, 25
  and bats 158–60, 161–2, 188–9
  and beavers 28–35
  and mines 169–74
  and mink 130–1
  and pine martens 77–8, 93–4, 101–2, 107
  and red squirrels 237–9, 263
  and seals 276, 278, 304
  and water voles 132, 133–4, 140
  and wild boar 39, 41–2, 56–7, 59, 74–5
  see also Boscastle (Cornwall); Forest of Dean (Glos); Formby (Merseyside); Horsey (Norfolk); Otter River (Devon); Scilly, Isles of
Environment Agency 16, 34, 145
Escot Park (Devon) 253–4
Eustice, George 15
Ewing, Fergus 293, 317
extinctions ix–x, xii, 48, 226–7, 316
  and bats 160
  and beavers 8, 20, 22–3
  and capercaillies 109

and pine martens 93
and wildcats 313, 315

faeces 89–91, 103–4, 176–7
farming, *see* agriculture
Farne Islands 290, 291
fashion 9, 256–7
feeding 209–10, 247, 248–9, 252
ferrets 83
field voles xiii, 103
Finland 8, 239–40, 254
fish stocks 15–19
fishing industry 284, 292–8
fleas 249
flood defences 11–12, 13, 14
Forest of Dean (Glos) 40–1, 42–3, 46, 49–50, 52–4, 57–8, 66
and bats 176
and butterflies 67
and pine martens 102, 112–13
Forestry Commission 43, 46, 55, 57–8, 240–1
and pine martens 113
and red squirrels 244, 263
Forestry England 65
forests 54–7, 102, 109; *see also* afforestation
Formby (Merseyside) 238–9, 246–9, 270–1
fossorial water voles 122–3, 125, 132, 137–8, 140–1, 144–5
foxes 6, 7, 104, 217–18
France 6, 60, 84, 143, 303–4
fur trade 86–7, 125, 126
and beavers 8–10, 27–8
and red squirrels 239–40, 256–7

Galloway Forest Park (Scotland) 99–100
Game and Wildlife Conservation Trust 108
Game of Thrones (TV series) 51
Garden Wildlife Health Project 213
geese 285–6
Georgia 71
Gerald of Wales 23–4, 27
Germany 6, 8, 28, 142, 241, 301
and bats 160
and hedgehogs 211
and pine martens 84, 85
and wild boar 57, 71, 73–4
ghost gear 284
Giardia duodenalis ('beaver fever') 5
Gillings, Hugh 114–15
'Give Seals Space' campaign 308
Glasgow 120, 121, 123, 137–9, 140–1, 145–51, 152–5
Global Biodiversity Information Facility (GBIF) 141
Goldsmith, Oliver 259
goshawks 53–4
Gove, Michael 211, 212
government xii, 92, 237
and beavers 3–4, 6, 13, 15, 16, 26, 31–2, 33–5
and grey seals 273, 290, 291, 292–3, 309, 310–11
and hedgehogs 193–5, 197, 199, 210–11, 212
and pine martens 96–8
and wild boar 38–40, 73–5
see also Defra
Gow, Derek 144–5, 146–7
grass 151–3, 285
Grayling, Chris 197
greater horseshoe bats 159, 160–9, 182–3, 190–2
and breeding 186–8
and diet 176–7
and genitalia 184–5
and hibernation 169–72, 174–6, 177–8
and lighting 178–80
and mating 180–1, 185–6
and roosts 188–9
Greenpeace 309–10
grey seals 272–4, 275–9, 282–3, 284–5, 301–7, 310–12
and breeding 286–8
and fish 295–8
and harbour seals 300–1
and protection 290–3, 307–9
and rescue pups 280–1
and water 288–90
Grey Seals (Protection) Act (1914) 291
grey squirrels 235–7, 238–9, 252, 264–5

and birth control 268–70
and culling 260–3
and Italy 263–4, 265–8
and pine martens 102, 103–7
and squirrelpox 248–51
Gurnell, John 250

Hafod Forest (Wales) 89–91
Haldane, J.B.S. 255–6
harbour seals 274, 275–6, 277, 291, 292, 298–301
Hardy, Thomas 162, 190
Harrower, Colin xii
hedgehogs xiii, 201–2, 217–19, 326
and badgers 199–201
and birdlife 222–6
and decline 193–6, 205–6
and diet 206–10
and infections 213–15
and Isles of Scilly 227–8, 233–4
and mating 216–17
and poisons 210–13
and Potter 204–5
and roads 196–9
and self-anointing 219–21
and Shakespeare 202–4
and taxonomy 215–16
Henry III of England, King 41
Heseltine, Michael 260, 261
hibernation 169–72, 174–6, 177–8, 206
Highland Squirrel Club 242–3
Homer: *Odyssey* 306
Horsey (Norfolk) 279–80, 281, 282–3, 284–8, 298, 311–12
Hungary 87
hunting 86, 239–40, 290–2, 311
and beavers 8–10, 23–5
and wild boar 41, 49, 51, 56–7, 71–3
Hurrell, H.G. 88
hybridisation 314
Hywel Dda, King 27–8

Iberian Peninsula 69, 84, 139, 143
insecticides 211–12
insects 176–7, 206, 207–9; *see also* butterflies
International Union for Conservation of Nature (IUCN) xii, 27, 40, 45–6, 158
and Red List 321, 324, 325
and red squirrels 239, 253
and seals 302
and water voles 122, 141
invasive species 130, 225, 235–6; *see also* grey squirrels; mink
Ireland 94, 102, 103, 241
and capercaillies 109
and grey squirrels 237
and red squirrels 256
Italy 60, 84, 144, 316
and Abruzzo National Park 318–19
and grey squirrels 263–4, 265–8

James I of England, King 57
Japan 59–60, 72, 215
Jefferies, Don 119
Jewitt, Llewellynn: *Grave-mounds and their Contents* 123–4
John Craven's Newsround (TV programme) 311
Johnson, Boris 33
Juvenal 23

Kashmiri goats 270
Keats, John 202, 258
kestrels 285
Knepp estate (West Sussex) 66–7
Kubasiewicz, Laura 79

Lacépède, Bernard-Germain-Étienne de la Ville-sur-Illon, Comte de 141
Lambert, Rob 311
Lambin, Xavier 129
lambs 50
Lancashire Wildlife Trust 247, 248
Land Reform Act (2003) 130
language 165–6, 305–6
Large Hadron Collider 85
Lewis, C.S.: *The Lion, the Witch and the Wardrobe* 15
lighting 178–80, 181, 210
Linnaeus, Carl 142–3, 183, 215
litters, *see* breeding

Longworth traps 232–3
Lord of the Flies (Golding) 48, 61
lungworm 213–14
Lyme Bay (Devon/Dorset) 310
lynx 45, 105, 316–17

Macbeth (Shakespeare) 203
Macdonald, Helen 54
Malta 160
Mammal Society xi, xii, 39, 135–6, 308
  and hedgehogs 197, 198
  and red squirrels 244, 257
Marine (Scotland) Act (2010) 291
Marine Conservation Zones (MCZs) 309
Marine Protected Areas (MPAs) 309–10
martens, *see* pine martens; stone martens
Martin, John 99
masting 245–6
mating:
  and bats 165–6, 180–1, 185–6
  and beavers 20
  and hedgehogs 216–17
  and pine martens 87
  and seals 277
  and wild boar 60
Matthews, Leonard Harrison: *British Mammals* 243
May, Brian 223
medicine 59, 257
Mendel, Gregor 142
metaldehyde 211
Midsummer Night's Dream, A (Shakespeare) 202
mink 83, 87, 104, 125–6, 135
  and controls 128–31
  and otters 127–8
  and water voles 119, 120–1, 137
moles 204
Monbiot, George 62, 316–17
Mongolia 257
monk seals 306
Montgaudier baton 303, 304
moose 104
moulting 257, 275, 277, 288, 306–7
Movement against Bats in Churches (MaBiC) 191
mustelids 82–3, 87

Nagel, Thomas 156, 157, 161
National Farmers' Union (NFU) 13, 32, 211, 317
National Trust 74–5, 246–7, 291, 311
Natural Environment and Rural Communities Act (2006) 96
Natural Environment Research Council (NERC) 262
Nature Conservancy 311
NatureScot 26, 27, 150, 224, 225–6, 313–14
Netherlands 70–1, 126, 160
New Zealand 216, 226
North Sea 298, 299, 308–9
Northern Ireland 105, 263
Norway 7–8, 26, 28, 293

offshore wind farms 308–9
O'Mahony, Kieran 74
OSPAR Convention 273
otariids 274–5
Otter River (Devon) 1, 4, 6, 7, 35–7
  and fish stocks 15–16, 17–19
otters 2, 3, 19, 83, 126–8
  and breeding 87
  and Red List 324–5, 327
Otterton (Devon) 2, 35–6
Ovid 51
Oxford, Earls of 56, 57

peasantry 56–7
People's Trust for Endangered Species 197
pesticides 210–11
pets 214, 244, 256
phocids 274–5
pigs 44, 45, 47–8, 66
  and disease 68–71, 93
  and swimming 61
pine martens 77–80, 83–4, 86–9, 98–100, 325
  and birds' eggs 113–14
  and capercaillies 110–12
  and diet 102–3

and reintroduction 80–2, 92–8, 100–2, 112–13, 114–17
and scats 89–91
and squirrels 103–7
pink-footed goose 285–6
pinnipeds 274
plastic 283–4
Plato 217–18
Pliny the Elder 218
Plitvice National Park (Croatia) 54–5
Poland 71, 72–3, 84, 107
and bats 79, 113
polecats x–xi, 83, 87, 89, 204, 326
Poliquin, Rachel 8
pork 69–70
porpoises 232, 273, 295, 298, 300
Portugal 316
possums 104
Potter, Beatrix 219, 220
The Tale of Mrs Tiggy-Winkle 204–5
The Tale of Squirrel Nutkin 258–9
predators 11, 91, 99, 103, 104–7
and bats 178–9
and capercaillies 110–12
and fish 295–7
and grey seals 300
Preston (Lancs) 251–2
public support 310–11
Pullman, Philip: *His Dark Materials* 88

rabbits xi, 54, 63, 125, 229, 257, 322
and decline 304
and disease 137
rabies 5
Rackham, Oliver 241
rat poison 212–13
Ratatoskr 259
rats 229
Ray, John 141, 183
Razorback (film) 51
Read's Island (Lincs) 132
Rebanks, James 63
Red List xi–xii, 119, 230–1, 321–32
and bats 160
and beavers 25, 27, 28–9
and hedgehogs 196
and wild boar 39, 40
red squirrels xii, xiii, 235–6, 237–40, 255–8
and captive breeding 253–4
and conservation 262–3
and decline 242–5
and Formby 246–7, 270–1
and habitat 252, 264–5
and Isles of Scilly 229
and mast years 245–6
and pine martens 102, 103, 104–5, 107
and Potter 258–9
and squirrelpox 247–50
and tree destruction 260
Red Squirrels United 263
Reeve, Nigel 201–2, 220, 221
reforesting, *see* afforestation
reintroductions 45, 86, 91–2, 317
and beavers 4–5, 7, 25, 27, 32, 34, 52
and otters 127–8
and pine martens 55, 80–1, 92–8, 100–2, 112–13, 114–17
and predators 104, 105–7
and red squirrels 242, 253
and water voles 136
and wild boar 57, 58
and wildcats 315
Review of the Population and Conservation Status of British Mammals, A xi–xii
rewilding 33, 66–7, 92, 94; *see also* reintroductions
Richard III (Shakespeare) 202
Richard of York 56
Ritchie, James 242
River Otter Beaver Trial 4, 31, 34
roads x–xi, 33, 89
and hedgehogs 196–9
and water voles 120, 135, 155
and wild boar 49, 73
Romania 316

St Mary's (Isles of Scilly) 226, 227–8, 231–4
salmon 16–19, 292–6
sand eels 225, 294, 299, 300
Sandom, Chris 64
Save our Beavers 4

scats 89–91, 103–4, 128
Schofield, Bridgit 82
Schofield, Henry 81–2
Scilly, Isles of xi, 226, 227–34, 279
Scotland 102, 161, 305
  and beavers 15, 22–3, 25–7
  and capercaillies 108–12
  and hedgehogs 223, 224–6
  and mink 129–30
  and pine martens 77–8, 79, 93–4, 99–100, 103, 107–8
  and red squirrels 237–8, 242–3, 253
  and salmon 293–6
  and seals 276, 280, 281, 291, 292–3, 298, 299, 301
  and water voles 124–5
  and wild boar 38–9, 41, 64
  and wildcats 313–14
  see also Glasgow
Scottish Gamekeepers Association 108, 111, 112
Scottish Wildcat Action 313–14
scrofula 59
sea lions 274–5
Sea Mammal Research Unit (SMRU) 273, 275, 284, 296
seals, *see* grey seals; harbour seals
seed dispersal 65–6
selkies 305
Shakespeare, William 51, 56–7, 202–4
sheep 95; *see also* lambs
shrews xi, xiii, 229–31, 232–3, 327
Simpson, Vic 127
Slater, David 52–3, 76
Spain 49, 72, 316
Special Areas of Conservation (SACs) 306, 307, 309
Special Committee on Seals (SCOS) 273
squirrelpox 235–6, 238–9, 247–51, 264
squirrels, *see* grey squirrels; red squirrels
Stewart, Robyn 140, 147, 155
Stewart, Rory 193–4
stoat xi, 83, 87, 204
Stocker, Les and Sue 201
stone martens 84–6, 113
Stourhead (Wilts) 74–5
Strachan, Rob 81, 131, 133
Sweden 8, 241, 316
swimming:
  and beavers 20–1, 29
  and seals 278–9, 304
  and water voles 131–2
  and wild boar 61, 73
swine fever 68–71, 72–4, 93

tapeworm 6–7
Tasmanian devil 227
taxonomy 141–4, 215–16, 244
teeth 134, 166, 217, 229–30, 324
  and beavers 19, 20, 21, 23
  and seals 286, 301
  see also tusks
Tempest, The (Shakespeare) 202–3
tenrecs 215–16
ticks 69
Tittensor, Andrew 243, 261
translocation, *see* reintroductions
tree-planting 136–7
trout 17, 18
tusks 47–8
Twm Sion Cati 114

United States of America (USA) 104, 105, 178
  and beavers 5, 8–10, 11, 14, 15, 16–17
  and pine martens 77, 78–9, 84
  and seals 292–3, 295, 304
  and squirrels 255, 260

vaccination 70
Vincent Wildlife Trust (VWT) x, 188, 315
  and pine martens 93–4, 95, 96–8, 99, 100–1, 102, 113
voles xiii, 6, 103; *see also* water voles
Vulnerable species 39, 40

Wales 39, 127, 270, 276
  and bats 161, 162
  and beavers 27–8, 34
  and grey squirrels 237
  and pine martens 77–8, 79, 89–91, 93–8, 100–1, 107

and red squirrels 237–8, 263
and water voles 135, 136–7
see also Devil's Bridge (Ceredigion)
walruses 228, 231–2, 274
warthogs 69
Warwick, Hugh 201–2
A Prickly Affair 224
water voles xii, 104, 118–25, 131–2, 133–7
and fossorial habits 144–5
and Glasgow 137–9, 140–1, 145–51, 152–5
and mink 128, 130
and taxonomy 141–2, 143–4
weasels 82, 83, 87
wetlands 10–15, 121, 135, 148
whales 111, 163, 175, 273, 305, 308
White, Gilbert 57, 123, 124
wild boar xiii, 38–47, 52–3, 57–62, 67–70, 74–6
and Abruzzo National Park 318
and characteristics 47–8
and dangers of 48–52
and diet 62–6
and disease 68–74
and hunting 56–7
and Poland 107
and Red List 324, 327
Wild Mammals (Protection) Act (1996) 205
wildcats 313–15, 319
WildCRU 133–4
Wildlife and Countryside Act (1981) 33, 75, 205, 243–4, 308
Wiltshire Bat Group 162, 170, 227
Wind in the Willows, The (Grahame) 155
Wolf River (Devon) 29–30
wolverine 83
wolves 104, 105, 107, 111, 318, 319
woodlands 42, 44, 56, 57, 62–3
and decline 240, 241–2
and pine martens 77–8, 82, 84, 98–100
and squirrels 264–5
see also Forest of Dean
Wright, Patrick 81–2

Yalden, Derek 242, 304
Yellowstone National Park (USA) 104, 105

zoonotic disease 5, 61–2, 68